T0376694

Groundwater and Climate Change

This book undertakes a scholarly assessment of the state of the art of law and policy perspectives on groundwater and climate change at the international, regional and national levels. A particular focus is given to India, which is the largest user of groundwater in the world, and where groundwater is the primary source of water for domestic and agricultural uses. The extremely rapid rise in groundwater use in many Indian states has led to a growing groundwater crisis that they must address. The existing regulatory framework has not adapted to the challenges and fails to address any environmental concerns. On climate change, India has adopted a policy framework that makes the link with water, but no legislation has followed up to make the link operational. The subject matter of this book has been widely debated with regard to each of its main two components separately. Bringing these two domains together is what makes this book unique. The link between climate change and groundwater has been acknowledged to some extent, and there is growing interest in studying the impacts of climate change on (ground)water. Similarly, in water and environmental law and policy, increasing attention has been given to the study of climate change and groundwater legal and policy frameworks but generally separately. This book contributes to filling this knowledge gap by drawing on contributions from leading experts in the field of environmental and water law and policy who have been involved in climate change and/or groundwater research.

The chapters in this book were originally published in a special issue of *Water International.*

Philippe Cullet is Professor of International and Environmental Law at SOAS University of London, UK, and a Senior Visiting Fellow at the Centre for Policy Research in New Delhi, India.

Raya Marina Stephan is an international consultant, expert in water law. She works in international projects related to water management and transboundary waters, with a special focus on groundwater.

Routledge Special Issues on Water Policy and Governance

https://www.routledge.com/series/WATER

Edited by: Cecilia Tortajada (IJWRD) – Institute of Water Policy, Lee Kuan Yew School of Public Policy, NUS, Singapore and James Nickum (WI) – International Water Resources Association, France

Most of the world's water problems, and their solutions, are directly related to policies and governance, both specific to water and in general. Two of the world's leading journals in this area, the *International Journal of Water Resources Development* and *Water International* (the official journal of the International Water Resources Association), contribute to this special issues series, aimed at disseminating new knowledge on the policy and governance of water resources to a very broad and diverse readership all over the world. The series should be of direct interest to all policy makers, professionals and lay readers concerned with obtaining the latest perspectives on addressing the world's many water issues.

Water Reuse Policies for Potable Use
Edited by Cecilia Tortajada and Choon Nam Ong

Hydrosocial Territories and Water Equity
Theory, Governance and Sites of Struggles
Edited by Rutgerd Boelens, Ben Crow, Jaime Hoogesteger, Flora E Lu, Erik Swynedouw and Jeroen Vos

Energy for Water
Regional Case Studies
Edited by Christopher Napoli

Legal Mechanisms for Water Resources in the Third Millennium
Select Papers from the IWRA XIV and XV World Water Congresses
Edited by Marcella Nanni, Stefano Burchi, Ariella D'Andrea and Gabriel Eckstein

Integrated Water Management in Canada
The Experience of Watershed Agencies
Edited by Dan Shrubsole, Dan Walters, Barbara Veale and Bruce Mitchell

Groundwater and Climate Change
Multi-Level Law and Policy Perspectives
Edited by Philippe Cullet and Raya Marina Stephan

Groundwater and Climate Change

Multi-Level Law and Policy Perspectives

Edited by
Philippe Cullet and Raya Marina Stephan

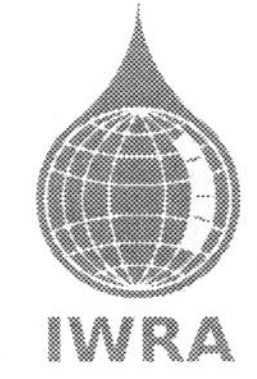

LONDON AND NEW YORK

First published 2019
by Routledge
2 Park Square, Milton Park, Abingdon, Oxon, OX14 4RN, UK

and by Routledge
711 Third Avenue, New York, NY 10017, USA

Routledge is an imprint of the Taylor & Francis Group, an informa business

British Library Cataloguing in Publication Data
A catalogue record for this book is available from the British Library

ISBN13: 978-1-138-48271-5

Typeset in Minion Pro
by codeMantra

Publisher's Note
The publisher accepts responsibility for any inconsistencies that may have arisen during the conversion of this book from journal articles to book chapters, namely the possible inclusion of journal terminology.

Contents

Citation Information

The chapters in this book were originally published in *Water International*, volume 42, issue 6 (August 2017). When citing this material, please use the original page numbering for each article, as follows:

Chapter 1

Introduction to 'Groundwater and Climate Change: Multi-level Law and Policy Perspectives'
Philippe Cullet and Raya Marina Stephan
Water International, volume 42, issue 6 (August 2017) pp. 641–645

Chapter 2

Regulating the interactions between climate change and groundwater: lessons from India
Philippe Cullet, Lovleen Bhullar and Sujith Koonan
Water International, volume 42, issue 6 (August 2017) pp. 646–662

Chapter 3

Assessing India's drip-irrigation boom: efficiency, climate change and groundwater policy
Trevor Birkenholtz
Water International, volume 42, issue 6 (August 2017) pp. 663–677

Chapter 4

Climate change, groundwater and the law: exploring the connections in South Africa
Michael Kidd
Water International, volume 42, issue 6 (August 2017) pp. 678–690

Chapter 5

Groundwater law, abstraction, and responding to climate change: assessing recent law reforms in British Columbia and England
Birsha Ohdedar
Water International, volume 42, issue 6 (August 2017) pp. 691–708

Chapter 6
EU legal protection for ecologically significant groundwater in the context of climate change vulnerability
Owen McIntyre
Water International, volume 42, issue 6 (August 2017) pp. 709–724

Chapter 7
Groundwater use in North Africa as a cautionary tale for climate change adaptation
Marcel Kuper, Hichem Amichi and Pierre-Louis Mayaux
Water International, volume 42, issue 6 (August 2017) pp. 725–740

Chapter 8
Global climate change and global groundwater law: their independent and pluralistic evolution and potential challenges
Joyeeta Gupta and Kirstin Conti
Water International, volume 42, issue 6 (August 2017) pp. 741–756

Chapter 9
Climate change considerations under international groundwater law
Raya Marina Stephan
Water International, volume 42, issue 6 (August 2017) pp. 757–772

Notes on Contributors

Hichem Amichi is based at the National Research Institute of Science and Technology for Environment and Agriculture (IRSTEA/UMR G-EAU), Montpellier, France.

Lovleen Bhullar is Senior Teaching Fellow at the School of Law, SOAS University of London, UK.

Trevor Birkenholtz is Associate Professor at the Department of Geography and Geographic Information Science, University of Illinois at Urbana-Champaign, USA.

Kirstin Conti is based at the Department of Geography, Planning and International Development Studies, Amsterdam Institute for Social Science Research, University of Amsterdam, The Netherlands.

Philippe Cullet is Professor of International and Environmental Law at SOAS University of London, UK, and a Senior Visiting Fellow at the Centre for Policy Research in New Delhi, India.

Joyeeta Gupta is Professor of Environment and Development in the Global South, Amsterdam Institute for Social Science Research, University of Amsterdam, The Netherlands.

Michael Kidd is a Professor at the School of Law, University of KwaZulu-Natal, Pietermaritzburg, South Africa.

Sujith Koonan is Teaching Fellow at the School of Law, SOAS University of London, UK.

Marcel Kuper is based at Cirad/UMR G-EAU, Montpellier, France, and at the Agronomy and Veterinary Sciences Institute Hassan II, Rabat, Morocco.

Pierre-Louis Mayaux is based at Cirad/UMR G-EAU, Montpellier, France, and at EGE, Avenue Mohamed Ben Abdallah Regragui, Rabat, Morocco.

Owen McIntyre is Senior Lecturer at the School of Law, University College Cork, Ireland.

Birsha Ohdedar is a Member of the Law, Environment and Development Centre, SOAS University of London, UK.

Raya Marina Stephan is an international consultant, expert in water law. She works in international projects related to water management and transboundary waters, with a special focus on groundwater.

INTRODUCTION

Introduction to 'Groundwater and Climate Change: Multi-level Law and Policy Perspectives'

Philippe Cullet and Raya Marina Stephan

Groundwater use has been growing throughout the world in recent decades, notably in India (United Nations, 2015). This poses a challenge that is widely recognized, yet with few exceptions the regulation of groundwater remains dated and too often focused on use rather than protection. Protection measures, where they exist, are generally conceived at the level of individual landowners, and not at the wider scale of the aquifer (Cullet, 2014). The shortcomings of the existing regulatory regime have become more and more apparent, in particular in warm and dry areas, where increasing groundwater use is associated with falling water tables (Dellapenna, 2013). Climate change, through its impacts on the water cycle, has made the situation much worse, and this will likely only deteriorate with time. Without effective action, prospects for a turnaround are dim (Taylor et al., 2013).

From a regulatory perspective, the most salient problem is that international legal instruments governing climate change do not focus much on groundwater, or water in general. This results in limited scope to link climate change policy with the regulation of groundwater, while the bias towards regulating use has often relegated the broader environmental dimensions to the back seat. In light of the limited pledges made by countries in the Paris Agreement (2015), increasing water use in general, and growing reliance on groundwater for agriculture and drinking water, it is imperative to forge more specific links between groundwater regulation and climate change governance.

This special issue undertakes a scholarly assessment of the state of the art of law and policy perspectives on groundwater and climate change at the international, regional and national levels. It arises from papers presented at workshops organized within a partnership project between the Law, Environment and Development Centre of SOAS, University of London, and the National Law University Delhi under the auspices of the UK-India Education and Research Initiative. This initiative sought to consider the impacts of climate change on groundwater, and to analyze and suggest corresponding improvements in the law and policy framework (SOAS, n.d.).

The partnership between SOAS and National Law University Delhi was thus set up in part to address the specificities of India's situation and to better understand how groundwater and climate change regulation can be linked by studying the situation in other countries and at the international level. India is the largest user of groundwater in the world. Groundwater is the primary source of water for its domestic and agricultural uses (Planning Commission, 2012). The extremely rapid rise in groundwater use in many Indian states has led to a growing groundwater crisis that they must address. The existing regulatory framework has not adapted to the challenges and fails to address any environmental concerns (Cullet, 2014). On climate change, India has adopted a policy framework that makes the link with water, but no legislation has followed up to make the link operational (Government of India, 2009).

At the same time, the challenges faced by India are relevant for many other regions of the world. It is thus crucial to examine the limited ways in which the link has been made and analyze how further progress can be made.

The subject matter of this special issue has been widely debated with regard to each of its main two components. The climate change regime has been the object of intense attention, first at the international level and progressively in most countries of the world (Carlarne, Gray, & Tarasofsky, 2016). Groundwater regulation has been the subject of increasing attention in the law and policy literature (Eckstein, 2017). Bringing these two domains together is what makes this special issue unique. It builds on existing scholarship in climate change by fostering a more direct engagement with water, and groundwater specifically. It similarly builds on existing scholarship on groundwater, which is often focused on issues of property rights when considered by lawyers and looked at from the perspective of a local resource by social scientists. In neither case does it give consideration to a global dimension.

The link between the two has been acknowledged to some extent, and there is growing interest in studying the impacts of climate change on (ground)water (Conway, 2013; Garner, 2016; Green, 2016). Similarly, in water and environmental law and policy, increasing attention has been given to the study of climate change and groundwater legal and policy frameworks, but generally separately. This special issue contributes to filling this knowledge gap by drawing on contributions from leading experts in the field of environmental and water law and policy who have been involved in climate change and/or groundwater research.

The articles gathered here address the legal frameworks governing groundwater and climate change. They consider how one domain deals with the other, at a country level, regionally, or globally.

The first two papers focus on India, a very large country. Cullet, Bhullar and Koonan (2017) note that the regime applied to groundwater in India contributes to the exclusion of the landless, while allowing individuals and companies with land and money to exploit the resource. According to them, the situation could be improved both for the people and the resource by integrating social equity and human rights dimensions in groundwater regulation. To do so would encourage the linking of climate change and groundwater. Similarly, viewing groundwater through the lens of climate change would give the opportunity to introduce measures of protection and conservation.

Birkenholtz (2017) questions the efficiency of drip irrigation technology in alleviating pressure on groundwater resources and adapting to climate change. In the state of Rajasthan in India, drip irrigation is being promoted to reduce groundwater abstraction and intensify agriculture, while no regulatory framework governs the use of groundwater. Farmers are extending the area under cultivation. The result is increased pressure on the over-exploited groundwater, which is also threatened by climate change. Birkenholtz concludes that promoting the adoption of drip irrigation technology in the absence of any rules governing groundwater, and with no consideration of ecological or social resilience, leads in the long term to further groundwater mining.[1]

Kidd (2017) looks at the case of South Africa, where groundwater is likely to be affected by climate change and other threats, such as acid mine drainage, pervasive water pollution and planned hydraulic fracturing. He sees the existing legal provisions as capable of addressing these risks. However, the low level of their implementation is often alarming. There is an important gap in the data required for appropriate decisions to be taken under the existing National Water Act that needs to be filled (see also Esterhuyse, Redelinghuys, & Kemp, 2016).

In his article, Ohdedar (2017) compares the recent law reforms in British Columbia and England with regard to water and groundwater. Both jurisdictions have taken steps to address the challenges of a changing climate. British Columbia enacted the 2014 Water Sustainability Act and introduced licensing for groundwater users and the payment of a fee except for domestic wells, while maintaining allocation based on the old system of first in time, first in right. England has adopted reforms to its legislation based on the EU Water Framework Directive, and introduced a number of innovations. In both cases, the legal framework requires a responsive regulator who can and will tackle issues of energy and fracking, which can exacerbate groundwater depletion and degradation.

McIntyre (2017) demonstrates how groundwater in the EU has received protection from the EU Habitats Directive. The Court of Justice has adopted a very strict interpretation of the maintenance of the 'integrity' of protected ecosystems, requiring precautionary assessment of groundwater's role. Its jurisprudence has involved robust protection of the key ecological characteristics and features of such sites, and offers an additional opportunity to protect groundwater and help sustain ecosystems and their services. The approach represents an informal step towards the integration of climate change adaptation measures into existing regulatory frameworks.

Kuper, Amichi and Mayaux (2017) observe the situation in North Africa, with a focus on Algeria and Morocco, where agricultural intensification has increased the pressure on groundwater. They conclude that currently states are unlikely to move to more sustainable policies to address either climate change adaptation or groundwater use. The authors have identified three conditions for the application of new governance schemes: obtaining wide legitimacy among the farmers and the rural areas; redirection of existing public investments; and engagement of a strong and active civil society pushing for more environmental policies.

The last two articles deal with the legal frameworks of climate change and groundwater at the global level. Gupta and Conti (2017) address the international climate change regime and global groundwater laws and show how they have developed independently. They observe other contradictions, such as differences in framing problems, in key norms and in processes. They provide recommendations to enhance the linkages between the two systems, such as close collaboration of institutions at the global level, and the integration of water in IPCC reports.

Finally, Stephan (2017) further considers both the international legal regime of climate change and transboundary aquifers. She investigates how the global climate change regime deals with groundwater, and similarly how international groundwater law addresses climate change. She concludes that water, and thus groundwater, receives more attention under the climate change framework than it appears because of due consideration being given to elements relying heavily on water, such as ecosystems and food production. These elements all fall under the adaptation regime. Cooperation on a transboundary aquifer for an adaptation plan following its specific process is fully in accordance with the international legal regime of transboundary aquifers.

Overall, the articles in this special issue show that climate change and groundwater have continued to be considered, and eventually regulated, separately, with little consideration of one by the other, whether at the national or the global level. This is so despite the strong physical link between groundwater and climate change. For example, groundwater can be affected deleteriously by climate change, but it also can act as a buffer against that change. Yet the present focus on regulating use does not take either effect into account. A better consideration of climate change with require more attention to environmental principles as well as a strong regulator for proper implementation, and a good knowledge of the resource. The climate change regimes, following international developments, have focused more on emissions. The Paris Agreement (2015) has opened a new approach by giving a strong position to adaptation, under which many water concerns fall. We hope that the implementation of this regime, to be further defined in a forthcoming Conference of the Parties, will lead to groundwater being given a much more visible place in recognition of its crucial role in climate change adaptation and mitigation in both the short and long term.

Notes

1. The phenomenon of farmers using 'saved' water to extend their irrigated area has been noted elsewhere, including in developed economies such as California (e.g., Frederiksen & Allen, 2011).

Disclosure statement

No potential conflict of interest was reported by the authors.

References

Birkenholtz, T. (2017). Assessing India's drip-irrigation boom: Efficiency, climate change and groundwater policy. *Water International, 42*(6). doi:10.1080/02508060.2017.1351910

Carlarne, C., Gray, K., & Tarasofsky, R. (Eds.). (2016). *The Oxford handbook of international climate change law*. Oxford: Oxford University Press.

Conway, D. (2013). Securing water in a changing climate. In B. Lankford, K. Bakker, M. Zeitoun, & D. Conway (Eds.), *Water security: Principles, perspectives and practices* (pp. 80–100). Abingdon: Routledge.

Cullet, P. (2014). Groundwater law in India: Towards a framework ensuring equitable access and aquifer protection. *Journal of Environmental Law, 26*(1), 55–81. doi:10.1093/jel/eqt031

Cullet, P., Bhullar, L., & Koonan, S. (2017). Regulating the interactions between climate change and groundwater: Lessons from India. *Water International, 42*(6). doi:10.1080/02508060.2017.1351056

Dellapenna, J. (2013). A primer on groundwater law. *Idaho Law Review, 49*, 265–323.

Eckstein, G. (2017). *The international law of transboundary groundwater resources*. Abingdon: Routledge.

Esterhuyse, S., Redelinghuys, N., & Kemp, M. (2016). Unconventional oil and gas extraction in South Africa: Water linkages within the population-environment-development nexus and its policy implications. *Water International, 41*(3), 409–425. doi:10.1080/02508060.2016.1129725

Frederiksen, H. D., & Allen, R. G. (2011). A common basis for analysis, evaluation and comparison of offstream water uses. *Water International, 36*(3), 266–282. doi:10.1080/02508060.2011.580449

Garner, E. L. (2016). Adapting water laws to increasing demand and a changing climate. *Water International, 41*(6), 883–899. doi:10.1080/02508060.2016.1214775

Government of India. (2009). *National water mission under national action plan on climate change*. New Delhi: Ministry of Water Resources.

Green, T. R. (2016). Linking climate change and groundwater. In A. J. Jakeman, O. Barreteau, R. J. Hunt, J.-D. Rinaudo, & A. Ross (Eds.), *Integrated groundwater management – Concepts, approaches and challenges* (pp. 97–141). Springer.

Gupta, J., & Conti, K. (2017). Global climate change and global groundwater law: Their independent and pluralistic evolution and potential challenges.*Water International, 42*(6). doi:10.1080/02508060.2017.1354415

Kidd, M. (2017). Climate change, groundwater and the law: Exploring the connections in South Africa. *Water International, 42*(6). doi:10.1080/02508060.2017.1351057

Kuper, M., Amichi, H., & Mayaux, P. L. (2017). Groundwater use in North Africa as a cautionary tale for climate change adaptation. *Water International, 42*(6). doi:10.1080/02508060.2017.1351058

McIntyre, O. (2017). EU legal protection for ecologically significant groundwater in the context of climate change vulnerability. *Water International, 42*(6). doi:10.1080/02508060.2017.1351060

Ohdedar, B. (2017). Groundwater law, abstraction, and responding to climate change: Assessing recent law reforms in British Columbia and England. *Water International, 42*(6). doi:10.1080/02508060.2017.1351059

Paris Agreement. (2015, December 15). UNFCCC, Decision 1/CP.21, Adoption of the Paris Agreement, UN Doc FCCC/CP/2015/10/Add.1, Annex. Retrieved 7 August 2017 from HYPERLINK https://protect-us.mimecast.com/s/oXA1BRUxao18t9?domain=unfccc.int http://unfccc.int/files/essential_background/convention/application/pdf/english_paris_agreement.pdf

Planning Commission. (2012). *Twelfth five year plan (2012–2017): Faster, more inclusive and sustainable growth* (Vol. 1). New Delhi: Government of India.

SOAS. (n.d.). UKIERI - Climate change and groundwater management in India (2013-2015). Retrieved from https://www.soas.ac.uk/ledc/research/ukieri—climate-change-andgroundwater-management-in-india-2013-2015.html.

Stephan, R. M. (2017). Climate change considerations under international groundwater law. *Water International, 42*(6). doi:10.1080/02508060.2017.1351911

Taylor, R. G., Scanlon, B., Döll, P., Rodell, M., van Beek, R., Wada, Y., … Treidel, H. (2013). Ground water and climate change. *Nature Climate Change, 3*, 322–329. doi:10.1038/nclimate1744

United Nations. (2015). *World water development report: Water for a sustainable world*. Paris: Author.

Regulating the interactions between climate change and groundwater: lessons from India

Philippe Cullet, Lovleen Bhullar and Sujith Koonan

ABSTRACT

Groundwater is often considered a largely local issue that is difficult to regulate. Further, groundwater regulation has often focused on use, rather than protection and conservation. There has thus been little integration of environmental concerns into groundwater regulation. Climate change calls for rethinking the regulatory framework for protecting and regulating groundwater. In India, the climate change regime has not given groundwater adequate prominence. Conversely, groundwater regulation remains largely detached from environmental challenges, including climate change. This needs to be addressed through regulation that links the two fields and is based on legal principles derived from the Constitution of India.

Introduction

Traditionally, surface water has been the primary source of water for different uses in India. The use of groundwater was restricted, partly due to technological constraints. However, the situation has undergone a dramatic change, in particular since the 1960s, and the demand for and use of groundwater, either as a supplement or as an alternative to surface water, have exponentially increased (World Bank, 2010). Some of the reasons relate to the characteristics of surface water resources, such as their seasonality and unreliability. Others concern groundwater resources, such as their ability to act as a resilient buffer against cyclical natural climatic variability, their quality, availability of low-cost abstraction technology, and the availability of free or subsidized electricity.

India is now home to the largest number of groundwater users in the world (WWAP, 2012). The problem is that the staggering increase in demand, linked to a rapid rise in the number of groundwater extraction structures, has led to a situation where use is often greater than recharge. The overall stage of groundwater development, or annual consumption compared to recharge, of the country is 62%. More specifically, 1071 over-exploited units, 217 critical units, 697 semi-critical units, 4530 safe units and 92 saline units were identified across the country in a study of 6607 assessment units (CGWB, 2014, p. ii).[1] This state of affairs highlights the urgent need for groundwater regulation. However, while surface water has been a subject of statutory regulation for a long time (notably in the form of irrigation laws), concerns over the quantity and quality of groundwater have received limited attention.[2]

The challenges posed by climate change in the water sector merit urgent attention at the international, regional, national and local levels. Water resources in a country like India are particularly vulnerable to the adverse impacts of climate change, including severe drought in some areas of the country and severe flooding in other parts. Therefore, the negative consequences of anthropogenic climate change provide the justification for the development of a new paradigm of groundwater regulation. On the one hand, groundwater has the potential to play a critical role in alleviating the adverse impacts of climate change on water resources by meeting future demand for different uses. On the other hand, climate change could adversely affect the quantity and quality of the available groundwater.

As a preliminary step towards the development of such a new paradigm, this article evaluates the scope and limits of the existing domestic regulatory framework in addressing the link between groundwater and climate change in India. The rest of the article is structured as follows. The section following this introduction examines the interactions between global climate change and groundwater in the specific case of India. In the absence of a comprehensive framework governing climate change and groundwater, the next two sections seek to unpack the domestic law and policy responses – first, the consideration of climate change (or the environment more generally) in the framework for groundwater, and second, the extent to which concerns relating to groundwater have been accommodated in the framework concerning the environment (generally) and climate change (specifically). This is followed by an examination of the need for a comprehensive framework to ensure the integration of climate change and more broadly environmental concerns in groundwater regulation.

Interactions between global climate change and groundwater in India

There are pervasive links between climate change and groundwater that impact the water cycle from the local to the global level. On the one hand, climate change is largely addressed at the global level, and the global water cycle is indeed directly linked to and impacted by climate change. On the other hand, groundwater is most often managed and regulated at the local level, since in most cases it is accessed directly from the aquifer and used locally. Yet, climate change directly impacts groundwater as the rainfall that recharges most aquifers becomes increasingly uncertain.

Both climate change and groundwater are of central importance in India. Groundwater is the source of about 80% of drinking water (Planning Commission of India, 2012a, p. 145),[3] accounts for around 60% of irrigation water use and is the only source of irrigation for the poorest farmers (Mukherji, Rawat, & Shah, 2013; Vijay Shankar, Kulkarni, & Krishnan, 2011). It is thus a lifeline for most people, and this makes it a key concern in terms of addressing the links between climate change and the water sector, in terms of both adaptation and mitigation (WWAP, 2012).

There has been a tremendous increase in the use of groundwater for irrigation, rising from a net irrigated area of 6 million ha in 1951 to 38 million ha in 2007–08 (Planning Commission of India, 2011). Reasons for these changes include the increasing availability and affordability of mechanized pumping and the unreliability of surface irrigation where it is available. In addition, since groundwater extraction, including over-abstraction, does not have the immediate visible consequences that drying rivers have, state governments have found it more politically convenient to subsidize the cost of groundwater extraction, such as through the provision of subsidized or free electricity.

Groundwater use may have positive or negative impacts on the local climate. One of the specific negative impacts is the contribution of groundwater extraction to India's greenhouse gas emissions (Shah, 2009). Rising exploitation of groundwater also leads to depletion of aquifers and increases the vulnerability of dependent ecosystems to the adverse impacts of climate change (Government of India, 2008). At the same time, groundwater is far more resilient to the impacts of climate change than surface water (World Bank, 2010). As a result, it can play an important role in climate change adaptation and disaster risk reduction by providing an additional or alternative source of water for different uses if managed sustainably (Villholth, 2009).

The links between (ground) water and climate change are apparent at various levels. Some of these links have been acknowledged in official documents for some time. Thus, the government of India noted in 2004 that 'the projected climate change ... will adversely affect the water balance in different parts of India and quality of ground water along the coastal plains' (MoEF, 2004). A number of possible impacts of climate change on groundwater resources have been identified. These include

- Influence on groundwater recharge due to changes in components of the water cycle such as evaporation, precipitation and evapotranspiration;
- Increased intrusion of saltwater into coastal and island aquifers due to rising sea levels;
- Increase in frequency and severity of flood and drought events that may affect groundwater quality in alluvial aquifers; and
- Increase in rainfall intensity, which may result in increase in flood events, higher surface runoff and soil erosion, and possibly reduced recharge (MoEF, 2012; Panwar & Chakrapani, 2013).

On the whole, however, the impacts of climate change on surface water resources have received far more attention than those on groundwater. The Fifth Assessment Report of the Intergovernmental Panel on Climate Change (IPCC, 2014) acknowledges that insufficient attention has been given to the relationship between groundwater and climate change. One reason is the physical and institutional invisibility of groundwater, the latter being due to land ownership being the determinant of access to groundwater. Hence, its use has been overwhelmingly a private affair rather than being subject to government control (World Bank, 2010).

There are several reasons to focus specifically on the groundwater–climate change link in law and policy. First, dependence on groundwater is likely to grow as a result of the increasing unreliability of availability of surface water. Addressing environmental challenges linked to water thus requires special attention to groundwater to ensure that the resource is exploited sustainably. Second, where the quality of surface water is impaired and where climate change may exacerbate this status, groundwater is central because it is often less susceptible to contamination and pollution. Third, the impacts of climate change on groundwater are likely to affect the realization of the human right to water, since groundwater is the primary source of drinking water and water for other basic human needs. Given that (ground) water is also a central element for the realization of other human rights, in particular the rights to food, to health and to a healthy environment, these rights also need to be taken into consideration when addressing the link between climate change and groundwater.

(Ground) water law and climate change: towards tackling climate change implications

The legal framework governing groundwater in India is structured mostly around the common law right of landowners to access and control groundwater beneath their land (Cullet, 2014). The law recognizes the right of landowners to extract groundwater without regulatory oversight. Some states in India have adopted groundwater laws, most of which are closely based on the Model Bill to Regulate and Control the Development and Management of Groundwater, prepared by the Union Government in 2005. In most cases, this legislation applies to groundwater in general, but the measures taken to regulate groundwater are quite limited in scope (Koonan, 2009). Some early-mover states, such as Maharashtra and Karnataka, initially adopted legislation with a focus on drinking water but subsequently enacted broader groundwater legislation (Maharashtra Groundwater (Development and Management) Act, 2009, and Karnataka Groundwater (Regulation and Control of Development and Management) Act, 2011, respectively).

The sustainability and equity problems and shortcomings of this regime are well recorded (Srinivasan & Kulkarni, 2014). However, there is little analysis of the regime as it relates to the impacts of climate change. It is anticipated that groundwater overdraft may deepen in the future due to climate change. Given its comparatively low vulnerability to changing rainfall patterns, in terms of both quantity and quality, groundwater can make up for insufficient surface water supplies, at least for some years. However, this significant feature of groundwater has yet to inform the legal regime in India, which is more or less indifferent to climate change–induced challenges. Since almost all the existing state-level groundwater laws have been adopted in the last 15 years, lack of knowledge or awareness cannot be cited as a reason for this oversight. Lawmakers were apparently not apprised of the intricate linkage between groundwater and climate change. In fact, there is hardly any literature making the link, particularly from a legal point of view. Further, most of the states followed the convenient route of adopting the 2005 Model Groundwater Bill, discussed in the next section, which in turn was based on another model document drafted in 1970, when there was hardly any concern about the environment, let alone climate change. As a result, the existing legal regime is insensitive to, and unprepared to deal with, climate change–related challenges and implications.

The next section analyzes how, and to what extent, the legal regime for groundwater in India addresses climate change-related challenges. It also discusses a way forward by analyzing the potential of the proposed model groundwater law drafted by the government of India to overcome the shortcomings from the point of view of climate change.

Land-based groundwater rights, unsustainable use and equity implications

One of the impacts of climate change is water-related stress in terms of both quality and quantity, resulting, for instance, from changing precipitation patterns. Water allocation is a major challenge during a crisis and such a situation demands a proper mechanism to ensure equitable allocation and use of the available sources. This is particularly relevant in the context of groundwater in India because it is the major source of drinking water and therefore a key source for the realization of the fundamental human right to water, whose core content includes at least domestic uses and arguably some livelihood uses too. Thus, climate change challenges demand a legal framework based on the principle of equity and the human right to water to regulate allocation and use of groundwater.

The existing legal regime for groundwater does not accommodate the principle of social equity, which calls for ensuring that no person is deprived of sufficient access to groundwater because of other persons' uses and the human right to water. This is mainly because it is based on the common law notion that groundwater is part and parcel of the land lying above the resource. Groundwater law in India follows this common law logic and gives uncontrolled power to landowners to use groundwater extracted from their land (Srinivasan & Kulkarni, 2014). Even though a number of states have enacted separate groundwater laws since the turn of the century, these laws are invariably based on the model legislation first prepared by the Union Government in 1970 and recast in slightly updated forms several times until 2005. This model legislation is based on the recognition that there is looming physical scarcity brought about by the large-scale introduction of mechanized pumping. Measures proposed thus centre around regulating or prohibiting new groundwater extraction in areas that have been identified as being over-exploited. This does not affect existing rights and access to groundwater, even where this may have contributed to falling water tables.

The exclusive right of landowners, as recognized by the existing groundwater law, contravenes established legal principles such as the public trust doctrine, the human right to water and sustainable development, because these principles do not approve a system of private appropriation of groundwater on the basis of land rights (Vani, 2009). It is a visible contradiction in the Indian legal system that while these principles are well recognized, one of the critical sources of freshwater continues to be available only to landowners.

One of the central problems is that the legal regime excludes the landless and allows individuals and companies with land and money to exploit groundwater to the detriment of others. The impact of the existing law on the human right to water is critical because groundwater is the major source of freshwater for drinking and other domestic purposes in rural areas and often an important source of water in urban areas. In fact, the large-scale over-exploitation of groundwater, which has been facilitated by the common law rule, has already triggered a number of conflicts relating to groundwater use and allocation in India. For example, local communities in a few states are protesting against water-based industries such as Coca-Cola and Pepsi for depleting and contaminating their groundwater sources through over-exploitation (Bijoy, 2006; Bhaduri, 2014; Koonan, 2010).

The state groundwater laws address groundwater exploitation from a physical-scarcity point of view at a broad level; they shy away from abolishing the land-based groundwater right. These laws do not mandate that groundwater be allocated so as to realize equity and human rights. By implication, the state groundwater laws only protect the rights of landowners from infringement by other landowners. They do not ensure access for the landless and other vulnerable people. The problem with this approach is that it seeks to address an issue which has its roots in local, national and international factors without changing the focus of regulation that remains at the level of individual landowners. In other words, the current legal regime remains based on an atomized understanding of groundwater that considers its regulation exclusively at the level of individual plots of land. This does not provide the basis for aquifer-wide measures, and even less for introducing a climate change dimension to groundwater regulation.

Overall, the existing legal regime that allocates groundwater exclusively to landowners is unsustainable in a context where groundwater sources are fast depleting and are increasingly being contaminated. Such an approach will exacerbate inequity and the violation of the human right to water in a scenario of increased water vulnerability due to different reasons including climate change. The land-based groundwater right, therefore, needs to be abolished and replaced by a new legal framework that is sensitive to climate change–related challenges.

Inadequate focus on protection and the need for an aquifer-based approach

Groundwater is central to climate change adaptation strategies because it is a more resilient freshwater source than surface water, and therefore it must be protected and conserved. However, the existing legal regime for groundwater in India lacks a strong emphasis on resource protection, which adds to its insensitivity to climate change–related challenges. The regime needs to focus more on protection to deal with the crisis. The present approach, from a protection point of view, is piecemeal in nature as it mainly focuses on controlling groundwater extraction by landowners in places where over-extraction is already taking place without addressing issues related to recharge of the aquifer. Similarly, while rainwater harvesting can be a significant conservation measure, it is conceived in the existing laws in narrow terms. Thus, groundwater laws focus mostly on rooftop rainwater harvesting in urban areas. Rainwater harvesting has been made compulsory for certain large buildings in some legal instruments, such as the Bihar Groundwater (Regulation and Control of Development and Management) Act, 2006 (§ 18), and the Bangalore Water Supply and Sewerage Act, 1964 (§ 72A). Yet, rainwater harvesting must also take place in rural areas, where individuals, communities and the government have taken a multitude of initiatives over time, including through the building of check dams. This should not only be encouraged but be made part of conservation strategies in legal instruments. This has not yet been done, and in some situations the rural population has been restrained from building such structures under the guise of ensuring better water flows to large dams (High Court of Rajasthan, 2004).

The regulation of over-exploitation and measures such as rainwater harvesting found in existing groundwater laws are beneficial, but there are strong limitations to this micro-level, individualized approach. Protection measures need to be based on a framework that considers the aquifer as a system, including its recharge and discharge areas. This is regarded as the proper unit for the purpose of groundwater protection and regulation (Kulkarni & Vijay Shankar, 2009). It will establish a connection between different aquifers and the water cycle in general that is insufficiently factored into existing legal instruments. It further provides a basis for integrating the links between groundwater and surface water in recognition of the fact that they form inseparable parts of a cycle ranging from the local to the global level. There are thus direct links between global climate change and groundwater protection that must be taken into account in legal instruments.

The existing and impending groundwater crisis, therefore, demands a strong framework for protection not just for the present generation, but also in the interest of future generations. In fact, groundwater law needs to be significantly informed by the principles of environmental law, such as intra-generational and inter-generational equity. The integration of these principles also helps to view groundwater as a part of the environment as opposed to the present regulatory view that considers groundwater as similar to a bucket that can be filled and emptied. A strong emphasis on protection and conservation also means strict regulation of activities (for example, encroachment in the recharge zones) that affect the recharge and discharge of aquifers. Examples from Rajasthan highlight the impacts of mining activities on groundwater, specifically the large amounts of water that are pumped to facilitate exploration, actual mining, and processing of mined materials. The unscientific disposal of waste also impacts groundwater, for example by blocking recharge (Cullet, Bhullar, & Koonan, 2015).

A framework for groundwater regulation and protection also requires an effective institutional framework. The institutional mechanism set up under the existing legal regime in India is incapable of tackling the challenges posed by climate change. Groundwater governance is marked by a centralized, command-and-control approach that is particularly unsustainable in the context of groundwater because it overlooks the highly decentralized nature of groundwater uses and the challenge of controlling the actions of millions of natural and legal persons.

There is thus a need for a more viable and effective institutional mechanism based on the principles of decentralization and participatory, democratic decision making. Such an institutional framework is not only viable and effective but is also consonant with the principle of decentralization envisaged by the 73rd and 74th amendments (in 1992) to the Constitution of India, which contribute to the realization of procedural rights of democratic participation at the local level.

Model groundwater bill of 2016: towards a framework sensitive to climate change–related concerns

To address the shortcomings of the legal regime, the Planning Commission of India adopted the Model Bill for the Conservation, Protection and Regulation of

Groundwater, 2011 (Cullet, 2012; Planning Commission of India, 2012b).[4] This has now been updated by the Ministry of Water Resources, River Development & Ganga Rejuvenation and is known as the Model Groundwater (Sustainable Management) Act, 2016. Although the Act does not include any explicit reference to climate change, several of its provisions highlight the environmental aspects of groundwater and thus contribute to the process of making the groundwater laws in India sensitive to climate change–related challenges.

The Act introduces a number of key changes to the existing legal tools for and approaches to groundwater regulation and protection, including the institutional framework for management. It takes a broad perspective on groundwater regulation that includes conservation and protection measures. This includes recognition in the preamble of the links between regulation of groundwater and climate change. This approach is in contrast to existing laws, which focus only on regulating uses of groundwater resources. The objectives of the Act include: promotion of sustainable groundwater use without compromising the needs of future generations; adoption of an integrated approach to groundwater and surface water to ensure conjunctive use of these water resources; protection of ecosystems and their biological diversity; and reduction and prevention of groundwater pollution and degradation. The Act provides for the preparation and implementation of binding, aquifer-based use and protection plans, known as groundwater security plans. It further envisages the creation of groundwater protection zones for need-based regulation and protection of groundwater where specific protection measures need to be taken. A strong emphasis on protection and conservation is one of the important changes in the climate change context. Further, the Act explicitly makes a link between groundwater law and environmental law principles such as inter-generational equity and the precautionary principle.

Another important change introduced by the Act that is relevant in the climate change context is a departure from the traditional approach based on the artificial division between groundwater and surface water for regulation and protection purposes. The Act recognizes the unitary nature of water and the integration of surface water and groundwater. This is an important step towards recognizing water as a part of the environment and its link with the water cycle at different levels. This is taken up further in the Draft National Water Framework Bill, 2016, drafted alongside the Model Groundwater Act, 2016, which seeks to provide an overall umbrella framework that links surface water and groundwater regulation.

The Model Groundwater Act, 2016, also recognizes the fundamental right to water as one of the foundational principles of the legal regime for groundwater and makes the realization of the right one of its principal goals. It also recommends the abolition of the land-based groundwater right and replaces it with the principle that groundwater is a common heritage of the people, held in trust by the government. These are important steps to foster equity in the regulation of access to and use of groundwater. It also helps address the issue of over-exploitation by the powerful and the rich and its implications for the poor and the vulnerable.

The Act envisages management and regulation of groundwater at the local level and thus respects the decentralization principle as envisaged by the 73rd and 74th amendments to the Indian Constitution. It provides for the establishment of groundwater committees at various levels to exercise regulatory and management powers. For example, the Gram Panchayat Groundwater Sub-Committee is entrusted with the power to prepare the groundwater security plan for the purpose of conservation and regulation of groundwater.

Overall, the Act seeks to replace the existing legal regime for groundwater with a new one that is based on human rights and the principles of environmental law. This updating of the Model Bill for the Conservation, Protection and Regulation of Groundwater, 2011, in 2016 reinforces the message to all concerned actors that a new paradigm is necessary within which ample opportunities to include climate change concerns are available. However, the major challenge is to what extent different state governments will make use of this model legislation. In a context where the Constitution gives states the primary power to regulate water, it is for states to adopt legislation and adapt the model legislation to suit their needs and circumstances so that it provides an appropriate framework in the long term.

(Ground) water and environmental law

Environmental law in India has been developed as a separate body of law at least since the 1970s, and it has evolved over time to address emerging issues, for example waste management, environmental impact assessment and coastal zone regulation. However, it has failed to address the issue of climate change in general and the implications for water security in particular. At best, the issue of implications of climate change on (ground)water is addressed in environmental law in a rudimentary and diffused manner.

Groundwater in environmental law

Indian domestic environmental legislation incorporates several provisions relating to quantity and quality of groundwater. However, these laws do not specifically fill gaps in the existing water laws, and the consideration of climate change issues is at best implicit.

The Water (Prevention & Control of Pollution) Act, 1974 (WPCPA), is relevant insofar as its primary objective is to prevent and control water pollution, which includes groundwater pollution (the *quality* dimension). However, the coverage of the law is limited to point sources of pollution (that is, domestic and industrial effluents). Nor does it set any standards for water pollution caused by agricultural practices and runoff pollutant levels, nor are they monitored (Comptroller and Auditor General of India, 2012). Under the Environment (Protection) Act, 1986 (EPA), the central government, through the Ministry of Environment, Forest and Climate Change (earlier the Ministry of Environment & Forests, MoEF), is empowered to take measures for protection and improvement of the environment. This has led to the adoption of standards for the discharge of domestic and industrial effluents into water bodies, including groundwater. Further, in the exercise of its powers under the EPA, the MoEF adopted the Environmental Impact Assessment Notification, 2006, and the Coastal Regulation Zone Notification, 2011, both of which include provisions relating to treatment and discharge of effluents into groundwater.

The domestic environmental laws also accommodate concerns relating to the availability of groundwater (the *quantity* dimension). The use of treated, recycled municipal wastewater as an alternative source for irrigation is encouraged, which can reduce or prevent over-exploitation of groundwater (Government of India, 2008; MoEF, 2012; MoWR, 2011), and the importance of wetlands in groundwater recharge is recognized in the Wetlands (Conservation and Management) Rules, 2010. The Central Ground Water Authority, which has been constituted under the EPA to 'regulate and control development and management of groundwater resources in the country', adopts a command-and-control approach to regulate and control groundwater extraction.[5] For this purpose, it has issued a list of notified areas where permission for groundwater extraction through energized means is granted only for drinking water purposes and subject to certain conditions. In non-notified areas, the authority will 'consider' granting a No Objection Certificate for groundwater withdrawal by new or expanding industries or infrastructure projects based on specified criteria (CGWA, 2015).

A violation of any of the above-mentioned provisions can attract legal action under the relevant provisions of the WPCPA or the EPA. Despite these provisions, however, there is growing concern about the quantity and deteriorating quality of groundwater and water more generally. This suggests that the existing legal provisions have not been effectively implemented, have failed to address the existing issues, or are insufficient to address emerging issues.

In this context, it is often judicial decisions that have filled some of the gaps. For instance, over the years, widespread quarrying/mining has been undertaken in flagrant disregard of the above-mentioned legal provisions, which has resulted in harm to riverbeds and groundwater. This has led the National Green Tribunal (NGT) to restrain any mining activity in the country without obtaining environmental clearance from the concerned authority and license from the competent authorities (NGT, 2013b).[6] The NGT has also prohibited illegal and unauthorized sand and minerals mining without leave of the concerned authority on beaches and other coastal areas (NGT, 2013b). Following various judicial interventions and lobbying from pressure groups, the Environmental Impact Assessment Notification, 2006, was amended to include new clearance requirements for sand mining, to close loopholes that were apparent in the previous framework, which only required clearance for sand mining in areas of more than five hectares (MoEFCC, 2016).

Judicial interventions have also highlighted concerns that have received little or no attention in domestic environmental law. For instance, taking note of the problem of indiscriminate over-extraction of groundwater, the NGT has prohibited illegal and unauthorized boreholes in some cities (NGT, 2013c), groundwater extraction for any use at a construction site (NGT, 2013a) or for use in packaged drinking water units (NGT, 2013d), and issuance of new tube-well connections to farmers (NGT, 2014b). The NGT has also prohibited activities that restrict or prevent groundwater recharge, such as permanent constructions (including concrete pavements) in parks (NGT, 2014a) and planting of eucalyptus trees (NGT, 2014b).

Courts have also stepped in to direct the concerned government authorities to formulate policies to address the lacunae. This was the case where in the absence of regulation, the uncontrolled digging of boreholes and tube-wells and activities of water tankers were resulting in groundwater over-extraction (Anonymous, 2013).

At the general level, the Supreme Court has incorporated sustainable development, the precautionary principle, the polluter-pays principle and the public trust doctrine into domestic environmental jurisprudence (Supreme Court of India, 1996). The application of any or all of these principles and doctrines is imperative to manage the interactions between climate change and groundwater. In practice, however, the law and policy frameworks have been more amenable to the application of sustainable development and the polluter-pays principle to groundwater. There has been less reliance on the precautionary principle, which ought to be invoked especially in a context where actions based on 'precaution' become necessary in the face of uncertainties that are common to both climate change and groundwater. However, there is a reversal in this trend, especially in cases concerning groundwater pollution before the NGT. The National Green Tribunal Act, 2010 (§ 20), explicitly requires the NGT to apply each of these principles in making its decision. As for the public trust doctrine, while its application to surface water resources is well settled (Supreme Court of India, 1996), the same cannot be said for groundwater (High Court of Kerala, 2005). Nevertheless, some of the recently proposed laws include a provision for management of water (generally) and groundwater (particularly) as a common-pool resource held in public trust, which is a welcome development. These include the Model Groundwater Act, 2016 (§ 9), and the Draft National Water Framework Bill, 2016 (§ 4).

Groundwater in climate change policy

There is no climate change law at the Union/central or state level in India, but there is increasing activity in the sphere of domestic policy. There is explicit recognition of the link between climate change and groundwater, and the adoption of domestic measures to address the emerging issues is envisaged.

The release of the National Action Plan on Climate Change (NAPCC) in 2008 marked the commencement of concerted efforts by the government of India to address the impacts of climate change on various sectors (including water) through the adoption of mitigation and adaptation measures. This was a departure from the previous piecemeal approach, where different government departments and ministries were responsible for addressing specific aspects of climate change. The NAPCC lays down the principles and identifies the approaches to be adopted to address the impacts of climate change. The approaches include existing as well as proposed actions, which are elaborated in eight national missions: National Solar Mission, National Mission for Enhanced Energy Efficiency, National Mission on Sustainable Habitat, National Water Mission, National Mission for Sustaining the Himalayan Eco-system, National Mission for a Green India, National Mission for Sustainable Agriculture and National Mission on Strategic Knowledge for Climate Change.

Water is specifically recognized as a 'climate-sensitive sector' (Government of India, 2008, p. 1), and one of the eight missions under the NAPCC is the National Water Mission (MoWR, 2011). The main goals of the National Water Mission (all of which are applicable to groundwater) are to

- Develop a comprehensive water database in the public domain and assess the impact of climate change on the availability and quality of water resources;

- Promote citizen and state actions for water conservation, augmentation, and preservation;
- Focus attention on over-exploited areas;
- Increase water use efficiency by 20 per cent; and
- Promote basin level integrated water resources management (MoWR, 2011, p. 5).

The NAPCC acknowledges that several of these programmes are not new. For instance, the specific action points to focus attention on over-exploited areas include an intensive rainwater-harvesting and groundwater-recharge programme, as well as promotion of traditional systems of water conservation through implementation of programmes for repair, renovation and restoration of water bodies.

The National Water Mission also recognises the need for the enactment and enforcement of groundwater legislation. It also states that the mission 'will take into account' the National Water Policy, 2012, which recognizes the links with climate change and includes a section on 'Adaptation to Climate Change'.

Pursuant to the NAPCC, a number of states (Arunachal Pradesh, Haryana, Himachal Pradesh, Karnataka, Madhya Pradesh, Mizoram, Nagaland, Odisha, Rajasthan, Sikkim, Tripura and Uttarakhand) have prepared their State Action Plan on Climate Change, some of which include groundwater-specific provisions. For instance, Karnataka's plan proposes a groundwater cess from which a groundwater fund will be created to finance groundwater recharge schemes proposed by public and private project proponents.

The measures identified in the NAPCC perform two functions simultaneously. They promote India's development objectives and also seek to yield 'co-benefits' to effectively address climate change (Government of India, 2008, p. 2). In other words, measures that effectively address climate change but do not further the country's 'development' objectives do not fall within the purview of the national climate change policy. The problem is that the measures enumerated in the NAPCC do not seem to be directed or guided by a clear or consistent approach or framework (Dubash, Raghunandan, Sant, & Sreenivas, 2013). This makes it look like the NAPCC adopts a business-as-usual approach (Thakkar, 2012). The co-benefits framework has advantages, but it may also restrict the extent to which innovative protection measures may be introduced in the water sector.

The effectiveness of any law or policy framework for groundwater regulation is contingent on the availability of information on the quantity and quality of the groundwater resources as well as the impacts of climate change on them. The NAPCC cites the large uncertainties concerning the spatial and temporal magnitude of the impacts of climate change as the reason for not considering it desirable to 'design strategies exclusively for responding to climate change' (Government of India, 2008, p. 13). However, the government of India has recognized the urgent need for more and better information on the quantity and quality of groundwater resources and the likely impacts of climate change to formulate long-term adaptation measures. The National Monsoon Mission (2012–17) is tasked with the prediction of monsoon rainfall variability in all spatial and time scales.

The increasing salience of climate change in the domestic context is also demonstrated by the renaming of the MoEF as the Ministry of Environment, Forests and Climate Change. However, significant decision-making power over groundwater resources still rests with the Ministry of Water Resources, which highlights the importance of cooperation, coordination and integration.

Groundwater and climate change: an urgent issue to address comprehensively

The previous sections make it clear that the legal framework has yet to effectively address the links between climate change and groundwater. Several issues can be identified.

An important issue is that groundwater law in India remains, like most of water law, largely disconnected from environmental law. The basic framework governing groundwater focuses overwhelmingly on issues of access to and control over groundwater linked to land ownership. The framework thus addresses concerns of availability and use but not protection or conservation of the resource. Further, regulation is limited to individual landowners, and there is no aquifer-wide perspective in existing regulation. Current legislation based on the Groundwater Model Bill 1970/2005 is inappropriate insofar as it entirely fails to rethink the framework within which groundwater regulation is conceived. Thus, it neither incorporates an environmental perspective nor reflects any concern for the links between the local and the global water cycle. The Model Groundwater Act, 2016, seeks to address these shortcomings by adopting a framework that is centred around protection measures. Yet, at present it is only a conceptual framework that needs to be adopted at the state level and implemented. And even if it were adopted by states, the proposed framework would not address all the concerns raised here since it lacks a global perspective on groundwater. It introduces an aquifer-wide protection dimension to regulation but does not make the link to broader national-level or global phenomena. This can be explained by the fact that this is a regulation conceived at the state level. Nevertheless, groundwater regulation must take into consideration rainfall and the global water cycle as well.

To some extent, the protection focus of environmental law in India addresses some of the gaps in water law. Water is an integral part of the subject matter of environmental law in general. Further, the WPCPA covers at least some of the relevant aspects of water pollution. Yet, the environmental law framework in India remains lacking in various ways. There is, for instance, no comprehensive legislative framework on water conservation and protection. As a result, water is considered as part of other environmental protection measures, such as those related to forests, and sector-wide impacts are often overlooked.

Beyond this, environmental law in its present form is not geared towards addressing the global dimensions of water or other environmental issues. This does not mean that domestic law should address issues that need coordination at the international level. It means that modern environmental law or water law in the age of climate change must be based on an understanding of the broader phenomena that influence and impact the implementation of the measures adopted at the local level. Thus, in a context where groundwater availability is linked to recharge, which is itself linked to precipitation, failure to link the local to the global leads to the adoption of incomplete frameworks that will fail at some point.

The current context is thus one where environmental law is relatively more advanced than (ground) water law in linking the two fields. This is noteworthy, and groundwater law desperately needs to be given a fresh lease on life with the introduction of groundwater regulation that considers not only use but also (and primarily) protection (the environmental dimension).

The next step concerns the introduction of measures that take into account the intricate links that exist from the local to the global level. Some of these links are easy to identify. For instance, in the State of Rajasthan, the High Court banned the construction of anicuts (check dams) above a certain height throughout the state to ensure that a sufficient amount of surface water fills a dam meant primarily to provide drinking water to urban residents (High Court of Rajasthan, 2004). This had an impact on the groundwater recharge strategies in every locality, while the benefits accrued mostly to a limited number of people, including the drinking water needs of the capital city of Jaipur. Such measures may be problematic in terms of restricting local water conservation and use strategies. Further, the measures may be completely ineffective if they are not conceived with actual rainfall in mind. There is potential for contradictory outcomes in a drought year, when the lack of aquifer recharge may lead to the need for transfer of water for drinking purposes from other parts of the state or the country.

It is also necessary to integrate various broad principles of environmental law into (ground) water law. Further, a global perspective that recognizes the links between the various phenomena from the local to the global level is also required.

Conclusion

Groundwater security is threatened by ongoing climate change. Indeed, the imbalance between demand for, and supply of, groundwater is likely to worsen because of the impacts of climate change. This will require strong and potentially controversial measures.

At present, the law and policy framework is deficient when it comes to addressing the broader connections between groundwater and climate change. First, the existing water law and policy framework fails to squarely address environmental concerns generally and climate change specifically. This is true despite 20 years of 'water sector reforms', which are premised on 'water scarcity' conceived as an environmental issue. Second, despite clear recognition of fundamental rights to environment, water and sanitation, neither water legislation nor environmental legislation has clearly integrated a human rights dimension, which could encourage the linking of climate change and groundwater in domestic regulation. Third, there is a tendency to view climate change as an issue to be addressed at the macro level, thus reinforcing the focus on centralization in water law rather than following the constitutional mandate of decentralization, which is particularly relevant in the context of groundwater regulation at the local level. Fourth, there are also implementation challenges that must be overcome. These include inadequate governance, inappropriate policies that provide no clear priorities or directions to government agencies in their responsibilities, and very limited financial and human resources to manage groundwater resources and water-supply systems.

To view groundwater through the lens of climate change in a country like India provides a window of opportunity to conceive groundwater regulation in the context of the global water cycle, comprehensively integrate protection and conservation at the aquifer level, effectively link surface water and groundwater, and recognize that mitigation of and adaptation to climate change need to be regulated on the basis of the principle of subsidiarity/decentralization. While this is a difficult task, it is not impossible, and would be well worth the effort.

Notes

1. Units are at the block/*mandal*/*firka* level, a subdivision of districts in the administrative division of India.
2. For a sample of irrigation laws, see http://ielrc.org/water/doc_irrigation.php.
3. This can be compared with the United States, where groundwater accounts for only 20% of water use (Kenny et al., 2009).
4. The Planning Commission of India was a body of the government of India set up by a resolution of the government of India in March 1950. Its key functions included assessment of the material, capital and human resources of the country, investigation of the possibilities of augmenting the resources, and formulation of a plan for the most effective and balanced use of country's resources. On 1 January 2015, through a resolution by the government of India, the Planning Commission of India was replaced by a new institution, the NITI Aayog (National Institution for Transforming India).
5. The list of notified areas is available at http://cgwa-noc.gov.in/LandingPage/Areatype/ListNotifed.pdf.
6. The National Green Tribunal was set up by the 2010 National Green Tribunal Act for the effective and expeditious disposal of cases relating to environmental protection and has jurisdiction over all civil cases where a substantial question relating to environment in involved (§14).

Disclosure statement

No potential conflict of interest was reported by the authors.

References

Anonymous. (2013). HC: Formulate bore, tube well policy. *The Herald (Panjim)*, 29 November 2013. Retrieved from http://www.heraldgoa.in/Goa/The-Sunday-Roundtable/hc-formulate-bore-tube-well-policy/72731.html.

Bhaduri, A. (2014). People of a Semi-arid Rajasthan Village Battle Coca Cola. Retrieved from http://www.indiawaterportal.org/articles/people-semi-arid-rajasthan-village-battle-coca-cola.

Bijoy, C. R. (2006). Kerala's plachimada struggle: A narrative on water and governance right. *Economic & Political Weekly*, *41*(41), 4332–4339.

CGWA. (2015). *Guidelines/criteria for evaluation of proposals/requests for groundwater abstraction*. New Delhi: Government of India, Central Ground Water Authority.

CGWB. (2014). *Dynamic groundwater resource of India (as on 31st March 2011)*. Faridabad: Central Groundwater Board.

Comptroller and Auditor General of India. (2012). *Performance audit of water pollution in India*. Report No. 21 of 2011-12. New Delhi: Ministry of Environment & Forests.

Cullet, P. (2012). The groundwater model bill: Rethinking regulation for the primary source of water. *Economic & Political Weekly*, *47*(45), 40–47.

Cullet, P. (2014). Groundwater law in India: Towards a framework ensuring equitable access and aquifer protection. *Journal of Environmental Law*, *26*(1), 55–81. doi:10.1093/jel/eqt031

Cullet, P., Bhullar, L., & Koonan, S. (2015). Inter-sectoral water allocation and conflicts: Perspectives from Rajasthan. *Economic & Political Weekly*, *50*(34), 61–69.

Dubash, N. K., Raghunandan, D., Sant, G., & Sreenivas, A. (2013). Indian climate change policy: Exploring a co-benefits based approach. *Economic & Political Weekly, 48*(22), 47–61.

Government of India. (2008). *National action plan on climate change*. New Delhi: Author.

High Court of Kerala. (2005). *Hindustan Coca-Cola Beverages v Perumatty Grama Panchayat* 2005(2) KLT 554.

High Court of Rajasthan. (2004). *Abdul Rahman v State of Rajasthan* DB Civil Writ Petition No 1536/2003, High Court of Judicature for Rajasthan at Jodhpur, Judgment dated 2 August 2004.

Intergovernmental Panel on Climate Change. (2014). Climate Change 2014: Synthesis report. Contribution of Working Groups I, II and III to the Fifth Assessment Report of the Intergovernmental Panel on Climate Change. Geneva: Author.

Kenny, J. F., Barber, N. L., Hutson, S. S., Linsey, K. S., Lovelace, J. K., & Maupin, M. A. (2009). Estimated use of water in the United States in 2005. *U.S. Geological Survey Circular, 1344*.

Koonan, S. (2009). Legal regime governing groundwater. In P. Cullet, A. Gowlland-Gualtieri, R. Madhav, & U. Ramanathan (Eds.), *Water Law for the Twenty-First Century: National and International Aspects of Water Law Reform in India* (pp. 182–204). Abingdon: Routledge.

Koonan, S. (2010). Groundwater: Legal aspects of the plachimada dispute. In P. Cullet, A. Gowlland-Gualtieri, R. Madhav, & U. Ramanathan (Eds.), *Water governance in motion: Towards socially and environmentally sustainable water laws* (pp. 159–198). New Delhi: Cambridge University Press.

Kulkarni, H., & Vijay Shankar, P. S. (2009). Groundwater: Towards an aquifer management framework. *Economic & Political Weekly, 44*(6), 13–17.

MoEF. (2004). *India – initial national communication to the United Nations framework convention on climate change*. New Delhi: Government of India, Ministry of Environment & Forests.

MoEF. (2012). *India – second national communication to the United Nations framework convention on climate change*. New Delhi: Government of India, Ministry of Environment & Forests.

MoEFCC. (2016). Notification S.O. 141(E) dated 15 January 2016 (Amendments in the Environment Impact Assessment Notification, 2006), Ministry of Environment, Forest and Climate Change.

MoWR. (2011). *National water mission under national action plan on climate change*. Comprehensive Mission Document Volume I. New Delhi: Government of India, Ministry of Water Resources.

Mukherji, A., Rawat, S., & Shah, T. (2013). Major insights from India's minor irrigation censuses: 1986-87 to 2006-07. *Economic & Political Weekly, 48*(26–27), 115–124.

NGT. (2013a). *Vikrant Kumar Tongad v Union of India and Others* OA No 59 of 2012, National Green Tribunal (Principal Bench), Order dated 11 January 2013.

NGT. (2013b). *National Green Tribunal Bar Association v Ministry of Environment and Forests & Others* OA No 171 of 2013, National Green Tribunal (Principal Bench), Order dated 14 August 2013.

NGT. (2013c). *National Green Tribunal Bar Association v NCT of Delhi* OA No 108 of 2013, National Green Tribunal (Principal Bench), Order dated 4 September 2013.

NGT. (2013d). *Vikrant Kumar Tongad v Union of India* Application No. 59 of 2012, National Green Tribunal (Principal Bench), Order dated 28 February 2013.

NGT. (2014a). *Akash Vashishtha v Union of India & Others* OA No. 165 of 2013, National Green Tribunal (Principal Bench), Order dated 20 February 2014.

NGT. (2014b). *Safal Bharat Guru Parampara v State of Punjab and Others* OA No 9 of 2014, National Green Tribunal (Principal Bench), Order dated 5 March 2014.

Panwar, S., & Chakrapani, G. J. (2013). Climate change and its influence on groundwater resources. *Current Science, 105*(1), 37–45.

Planning Commission of India. (2011). *Report of the working group on sustainable groundwater management*. New Delhi: Government of India.

Planning Commission of India. (2012a). *Twelfth five year plan (2012–2017) – faster, more inclusive and sustainable growth – volume 1*. New Delhi: Government of India.

Planning Commission of India. (2012b). Model bill for the conservation, protection and regulation of groundwater 2011. In Planning Commission. *Report of the Steering Committee on Water Resources and Sanitation for Twelfth Five Year Plan*. New Delhi: Government of India.

Shah, T. (2009). Climate change and groundwater: India's opportunities for mitigation and adaptation. *Environmental Research Letters*, *4*(3), 1–14. doi:10.1088/1748-9326/4/3/035005

Srinivasan, V., & Kulkarni, S. (2014). Examining the emerging role of groundwater in water inequity in India. *Water International*, *39*(2), 172–186. doi:10.1080/02508060.2014.890998

Supreme Court of India. (1996). *MC Mehta v Kamal Nath* (1997) 1 SCC 388.

Thakkar, H. (2012). *Water sector options for India in a changing climate*. New Delhi: South Asia Network on Dams, Rivers and Peoples.

Vani, M. S. (2009). Groundwater law in India: A new approach. In R. Iyer (Ed.), *Water and the laws in India* (pp. 435–473). New Delhi: Sage.

Vijay Shankar, P. S., Kulkarni, H., & Krishnan, S. (2011). India's groundwater challenge and the way forward. *Economic & Political Weekly*, *46*(2), 37–45.

Villholth, K. G. (2009). The neglected role of groundwater in climate change adaptation and disaster risk reduction. *IOF Conf Series: Earth and Environmental Science*, *6*(29), 292062.

World Bank. (2010). *Deep wells and prudence: Towards pragmatic action for addressing groundwater overexploitation in India*. Washington, DC: Author.

WWAP. (2012). *The United Nations World Water Development Report 4: Managing water under uncertainty and risk*. Paris: UNESCO/WWF - World Water Assessment Programme.

Assessing India's drip-irrigation boom: efficiency, climate change and groundwater policy

Trevor Birkenholtz

ABSTRACT
This article draws on a case from the north-western Indian state of Rajasthan to examine whether drip irrigation saves water. Drip irrigation is being promoted to preserve groundwater and enhance resilience to climate change. However, the article finds that in the absence of regulations over groundwater abstraction, farmers acquire drip irrigation to intensify production rather than to conserve water. This occurs in a political and economic context where farmers are incentivized to do so, further exacerbating groundwater overdraft. The article concludes with a discussion of drip irrigation's impact on farmers' livelihoods and its implications for groundwater policy.

Introduction

Groundwater depletion is widespread in both semi-arid and humid regions of the world, while global climate change is predicted to exacerbate already dwindling groundwater aquifers by disrupting recharge rates (Aeschbach-Hertig & Gleeson, 2012; Taylor et al., 2013). This is particularly troubling for the future of groundwater-based irrigation, which currently accounts for 67% of groundwater withdrawals globally (IGRAC, 2010; Siebert et al., 2010). State water agencies around the world, as well as international water organizations, agree that the amount of groundwater withdrawn for irrigation must be reduced to reverse groundwater overdraft and to free up water for other uses. Yet given growing food security needs, this must happen while also raising agricultural productivity (Berbel & Mateos, 2014). Given that in many places it is politically contentious or even seemingly impossible to pass new groundwater-use regulations (Fishman, Devineni, & Raman, 2015; Narayanamoorthy, 2004), the focus on enhancing the efficiency of groundwater irrigation immediately through the deployment of new water-conserving technologies is taking centre stage.

Drip irrigation, which supplies water directly to plant stems or roots, is presented as a way to meet these objectives. Proponents have asserted that drip irrigation can double water-use efficiency compared to conventional, sprinkler irrigation systems (Worldwatch Institute, 2013). Over the past 20 years, drip irrigation systems have expanded sixfold globally (Worldwatch Institute, 2013) both through private initiatives by small and large farmers (IWMI, 2006) but also because of country-level subsidies, which incentivize their adoption by farmers (IWMI, 2006; Pullabhotla, Kumar, & Verma, 2012). Drip-irrigated area now totals over 10.3 million hectares across the world (National Geographic, 2013). Most of this growth has occurred in the arid and semi-arid regions of the United States, India and China, where there is often a primary reliance on groundwater for irrigation and drinking water needs. Today, India leads the world in both the rate of expansion in drip-irrigated area (111-fold over the past two decades) and in area under drip irrigation, with 2 million hectares (National Geographic, 2013). Indian state planners, development donor agencies, and scientists are promoting drip irrigation as a way to reduce groundwater withdrawals, enhance agricultural productivity and mitigate weather variability, while enhancing resilience to climate change (Government of India, 2010, 2014).

Yet while drip irrigation systems may enhance water-use efficiency, they may not actually save water (van der Kooij, Zwarteveen, Boesveld, & Kuper, 2013). For instance, Ward and Pulido-Velazquez (2008) found that the conversion of sprinkler irrigation systems to drip systems in North America actually increased water use via intensification, exacerbating groundwater decline and worsening soil conditions. While the effect of drip irrigation on groundwater and soil conditions is a matter of intense debate (Pei et al., 2015), farmers' rationale at the household level for whether to adopt drip irrigation has received little attention (but see Benouniche, Kuper, Hammani, & Boesveld, 2014) and would yield valuable insights as to the conditions under which the diffusion of drip-irrigation systems leads to more or less water use.

This article examines the multi-scalar political economy of drip-irrigation adoption by focusing on household decision making and government policy in Rajasthan, India. For the former, I rely on interviews with farmers who have recently adopted drip-irrigation systems to examine the specific conditions under which they have adopted these systems and with what goals. With respect to the latter, I examine recent agricultural policies of the government of India, as well as the government of Rajasthan, to understand the policy mechanisms that promote adoption by particular types of farmers.

Theoretically, the article draws on recent research from political ecology focused on irrigation efficiency that highlights the operation of the Jevons paradox (rebound effect) in water-conservation technologies, where more efficient technologies actually lead to greater use of natural resources (Dumont, Mayor, & López-Gunn, 2013). The article argues that gains in productivity as a result of the adoption of drip irrigation systems are not leading to less water use. This is because of the political economy of groundwater irrigation in India, which incentivizes intensification and extension of the irrigated area of commercial, high-water-demand crops. These findings highlight that water-conservation technologies are not politically neutral and will not lead to water conservation on their own or in the absence of new institutional arrangements (groundwater usage rules and laws).

The article proceeds in four further sections. The next section selectively surveys the literature on efficiency, both technical and economic, in groundwater-irrigated agriculture to show how this literature frames efficiency and groundwater abstraction in technical and apolitical terms. This dominant framing, the section argues, would benefit from a focus on the political economy of agricultural production and farmer motivations for the adoption of drip irrigation, both of which condition water use and potential savings. The third section introduces the detailed case study site of Rajasthan, India, by describing the social, physical and political-economic groundwater irrigation landscape, including how this is changing under climate change–induced weather variability. The fourth section draws on empirical research with farmers from 2015 and 2016 to demonstrate the political economy that incentivizes intensification of irrigation, undermining the dominant rationale that drip irrigation will lead to groundwater savings. The fifth section offers a discussion of the political economy of adoption and efficiency, and examines policy mechanisms that could be instituted for efficiency gains and lower water use, while enhancing irrigated agriculture's resilience to climate change. The final section concludes by detailing the lessons that can be drawn from these cases and how they inform water policy.

The political economy of efficiency and farmer adoption in drip irrigation

There is a global consensus among development donor agencies on the need for efficiency gains in irrigated agriculture (Perry, 2007; World Bank, 2014). Yet what we mean by efficiency and whether drip irrigation in the field leads to efficiency gains or not, is much debated (Boelens & Vos, 2012; Lankford, 2012; van der Kooij et al., 2013). Boelens and Vos (2012) identify two basic discussions over water-use efficiency in irrigation: 'technical irrigation efficiency' and 'economic water allocation efficiency' (see also Lankford, 2006, 2012; Molle, Venot, & Hassan, 2008; Perry, 2007). The first is focused on increasing the ratio of water taken up by plants to water applied, or the amount that is 'beneficially used'. In enhancing technical efficiency, the goal has historically been to minimize 'water losses' in conveyance (e.g., leakage, evaporation) in the case of canal irrigation and/or the over-application or misapplication of water in irrigation generally. This notion was popularized by former UN Secretary-General Kofi Annan's proclamation that considering water scarcity and food insecurity, the global water community must find ways to achieve "more crop per drop" in irrigated agriculture. Historically, this has largely been the domain of irrigation engineers, who attempt to engineer more technically efficient irrigation systems, such as drip or micro irrigation, and crop scientists or agronomists who attempt to breed seed cultivars that require less water, while producing more of any given commodity (e.g., grain, tomatoes).

Economic water allocation efficiency, on the other hand, is primarily the domain of neoclassical and/or agricultural economics, which focus on pricing water at its scarcity value by charging water users the full marginal cost of supplying the water (Johansson, Tsur, Roe, & Doukkali, 2002, cited in Boelens & Vos, 2012). The goal of achieving economically efficient irrigation systems with respect to allocation and water use is to, first, generate more economic surplus per unit of water used, and second, to ultimately (re)direct water to uses that produce higher marginal returns per unit of water using pricing mechanisms. Boelens and Vos (2012), drawing on a political ecology framework, identify three main policy problems that arise due to the focus on these two forms of efficiency as panaceas in irrigation. First, 'efficiency' discourses may promote policies and projects that deprive peasant farmers or smallholders of their water use rights (particularly if these are not codified, as in much of India). Second, water policy and project notions that are driven by water experts (water and/or irrigation engineers) and are based on these two forms of efficiency can undermine local water management practices and may undermine livelihood strategies and security. And third, by setting technical and economic efficiency goals that are often unachievable (i.e., attainable only in a controlled laboratory setting or in theory under assumptions of perfect knowledge and competition), smallholders are rendered underachievers according to the norms established by the 'dominant power-knowledge structures'. In this way, farmers themselves are rendered inefficient and their decision making irrational because these goals do not take into account the complex political ecologies (including political economic circumstances) within which farmers make cropping and irrigation decisions (Birkenholtz, 2009). So, for instance, when farmers elect to grow subsistence or fodder crops rather than market commodities, this is viewed as an inefficient allocation of water and therefore an economically irrational outcome, even though it may lead to less water use overall and enhance resiliency to weather perturbations. Indeed, farmers in India have been known to switch from commercial crops to subsistence varieties during times of water scarcity because non-commercial crops are used as fodder for dairy cattle, which produce the main source of protein (milk) for most of the Indian rural population (Birkenholtz, 2009).

I would add to the three critical points identified above that the focus on efficiency in technical and economic terms has two further negative effects. First, the focus on water use efficiency in these two ways renders the issue of reducing groundwater decline through efficiency enhancements in irrigation a technical problem, demanding technical solutions (Birkenholtz, 2014). Following Li (2007), who argues that development interventions tend to transform the domain of these interventions into problems that require technical solutions (what she has termed 'rendering technical'), development experts identify particular problems and connect them to specific solutions. Here it is important to note that 'technical solutions' does not necessarily mean a physical technology but refers to creation of a 'solution' that renders and narrows the domain of intervention with 'specifiable limits and particular characteristics' (Rose, 1999, p. 33). Applied to the case of groundwater irrigation, development and government experts identify a problem – underperforming irrigation systems along lines of technical and/or allocation efficiency – that can be corrected via a specific solution, in this case the dissemination of new drip-irrigation technologies, rather than considering the political economic conditions under which farmers engage in irrigated agriculture. In doing so, the political economy that incentivizes irrigation intensification via more groundwater extraction is rendered opaque (Benouniche et al., 2014). This allows blame for over-extraction to be placed on farmers – directly, due to their supposedly poor production decisions, or via secondary factors, such as electrical subsidies, which are viewed as the prime problem that incentivizes over-extraction (see Shah, Scott, Kishore, & Sharma, 2004 for a discussion). Here both forms of efficiency become apolitical aspirations. This is a particularly salient point in contexts with little or no regulations over groundwater use and/or with little promise of achieving them. This is the case in north-western India.

In these circumstances, adopting a political economy perspective redirects attention both to the need to understand farmer motivations in adopting drip-irrigation systems and the broader political economic context within which farmers are producing commodities. Through this perspective, the precise causal processes that lead to Jevons paradox may be known, which will yield insights on what kinds of new institutions or rules may incentivize farmers to reduce water withdrawals. The article returns to this in the conclusion. Next, we consider the context of groundwater irrigation in Rajasthan, focusing on its social, physical and political-economic character.

Groundwater, climate change and drip irrigation in India

According to Shah (2009), 88–91% of groundwater withdrawals throughout India go towards irrigation. Groundwater decline, due to extraction that exceeds recharge, mostly for irrigation, and its exacerbation due to climate change, is a serious concern. According to data from NASA's Gravity Recovery and Climate Experiment, north-western India had the highest groundwater depletion rates in the world in 2002–08, even though precipitation was above normal for the period (Tiwari, Wahr, & Swenson, 2009). The severe state of groundwater overdraft in north-western India will interact with and probably be exacerbated by climate change–induced perturbations in at least two ways. First, the IPCC predicts a 0.5–1.0 °C rise in average temperatures by 2029 and a 3.5–4.5 °C rise by 2090–99, though warming of up to 5.0 °C has been predicted for the region (Yadav, Kumar, & Rajeevan, 2010). This increase in temperature is expected to be accompanied by an increase in extreme events (drought and significant rainfall episodes), leading to an increase in runoff of up to 40% by 2090–99 (Gosain, Rao, & Basuray, 2006). It is predicted that much of this runoff will not lead to increases in aquifer recharge due to the intensity of the predicted events (where significant rainfall occurs over a short period), coupled with a lack of rainwater harvesting and recharge structures (though the government is trying to expand the number and scope of these structures; see Government of Rajasthan (GOR), 2010). Warmer temperatures, coupled with fewer and more sporadic precipitation events, will lead to both less groundwater recharge and, simultaneously, growth in demand for groundwater irrigation (Shah, 2009).

This severe state of groundwater decline and heavy dependence on groundwater for irrigation has led to calls for both groundwater regulation and enhancing irrigation efficiency throughout India and Rajasthan.

Groundwater in India: regulation and efficiency

The regulation of groundwater in India is the purview of individual states, but the central government does try to provide direction in this regard. The primary means through which this has occurred is via the drafting of 'model bills' at the central-government level and then encouraging states to adopt some form of groundwater regulation based on them. The Model Bill to Regulate and Control the Development and Management of Ground Water was first drafted in 1970. Since then, it has been revised five times, most recently in 2016 (Cullet, 2012, 2014). Various forms of it have been adopted by 13 (of 29) states, though Rajasthan is a notable absence. The low rate of regulatory implementation has led the central government to actively pursue and promote increased efficiency in irrigation as a way to address groundwater overdraft and promote agricultural resiliency in the face of climate change.

The National Action Plan on Climate Change, 2008, was a major step in this regard. The plan created eight new bodies, or missions, of which two focused directly on water and agriculture. The National Mission for Sustainable Agriculture, 2010, recognizes the 'risks to [the] Indian agriculture sector due to climatic variabilities and extreme events' and seeks to 'encourage adoption of technologies for enhancing water use efficiency'. The National Water Mission, 2011, calls for an 'increase in water use efficiency by 20%' to improve climate resilience. Here the action plan combines groundwater decline, climate variability, water-use efficiency and new efficiency-enhancing water technologies as the future of sustainable agriculture and resiliency. Most recently, these goals will be embodied in the new National Bureau of Water Use Efficiency, the establishment of which is expected in the near future (Parsai, 2013). But it must be noted that, while the spectre of climate change is prompting this latest effort, the Indian central government has been actively pursuing drip irrigation as a means to enhance technical efficiency in irrigation since at least the 1980s (see Narayanamoorthy, 2010, for a discussion of the case of Maharashtra, India).

These efforts by the central government mirror the 'global consensus' on the need for efficiency, noted previously (and critiqued) by Boelens and Vos (2012). Under the National Mission for Sustainable Agriculture, 2010, the central government has allocated US$ 235 million (US$ 124 million in 2014–15 alone), targeted directly at encouraging the adoption of micro (including drip) irrigation. Today, micro irrigation has become so important to groundwater and agricultural policy that it now has its own separate government mission: The National Mission on Micro Irrigation (NMMI). The NMMI states that India's micro-irrigation potential is 69 million hectares, and its goal is to fully achieve this potential (up from the current 2 million ha), though a target date has not been specified. Interestingly, the stated goals of the NMMI are more focused on increasing area under drip and micro irrigation than reducing water demand in irrigated agriculture. Yet this remains the goal of the umbrella missions under which the NMMI is operating.

The NMMI allocates funding to individual states on a 50:10:40 basis, where the central government pays 50% of the implementation cost, while state governments and individual 'beneficiary' farmers pay 10% and 40%, respectively. The subsidy packages are implemented by the states, but through a devolved administrative structure that involves the funds making their way through a maze of bureaucratic entities, beginning with the State Level Micro Irrigation Committee and ending five layers down with the village-level *panchayati raj* elected government institutions and local private irrigation dealers. Individual states also have much discretion in how they implement the programmes (Government of India, 2014). Though implementation of the programme continues to evolve almost seasonally, states may add additional subsidies, offer different levels of training and maintenance support, offer loan guarantees and alter the contract process to make it more accessible to the smallest farmers (see also case studies by Kuppannan, Mohan, Kakumanu, & Raman, 2011; Pullabhotla et al., 2012). The subsidies fund 'drip irrigation system' packages (see below) sized by area in 1-hectare increments (up to a maximum of 5 ha per farmer) based on individual farmer applications.

Like much of India, Rajasthan is aggressively pursuing efforts to enhance irrigation efficiency by encouraging the diffusion of drip-irrigation systems. The next section details the socio-ecological context in which this is occurring, while highlighting the evolution of government incentive programmes with farmers' adoption efforts.

Groundwater in Rajasthan: regulation, policy and efficiency

Rajasthan is arid and semi-arid, highly socially differentiated along lines of caste, class and gender, and heavily dependent on groundwater. Indeed, groundwater serves 76% of Rajasthan's irrigated area – 4.4 million hectares – and 80% of domestic water supply (Government of India (GOI), 2014). Groundwater irrigation throughout the state is composed almost entirely of small-scale private farmers with landholdings that range from 0.5 to 20-plus hectares (GOR, 2011). Rajasthan historically has no formal state regulation that limits groundwater pumping or construction wells, including tube-wells (though this is changing, with construction permits now required in some zones). This has led to rapid groundwater decline in the state, where extraction exceeds recharge by at least 410 million m^3 per year (Government of Rajasthan Groundwater Board, personal communication, January 2006).

After years of failed attempts at adopting a version of the central government's Draft Groundwater Bill, the state of Rajasthan finally enacted the Rajasthan State Water Policy, 2010. The policy has at least three features salient for the present article. First, it calls for a reduction in water withdrawals for irrigation from 83% of total water withdrawals to 70%. Second, it seeks to 'optimize water' usage via new water-conservation technologies, stating that 'groundwater will be better utilized ... by facilitating drip irrigation techniques' (GOR, 2010). And third, it calls for greater participation of the private sector in facilitating technical and economic efficiency gains in irrigation and water use (both in irrigation and in other sectors). These three goals are reiterated in a recent report by the World Bank Group's International Finance Corporation, which sees an opportunity in Rajasthan's groundwater decline and irrigation inefficiency, understood as a technical problem: 'Agriculture, which consumes 83% of the state's water resources, presents the single biggest opportunity for water reform [including] a range of water savings interventions by the private sector' (Hooda, 2014, p. 10). Indeed, drip-irrigation design and manufacturing firms such as Jain Irrigation and Netafim-India are working closely with the government of Rajasthan to help farmers apply for state subsidies, and to provide setup and installation expertise. Yet farmers may have objectives other than saving water in using drip irrigation, as will be shown in the following section, which draws on recent qualitative field research.

Farmers' rationale for using drip irrigation: intensification and extending cultivation

The author has been conducting research on conventional irrigation in Rajasthan since 2001 and first noted the existence of drip irrigation on commercial seed farms in the mid-2000s and on non-seed farms by commodity farmers in 2011. The present article relies on in-depth and repeated interviews with 12 drip-irrigating farmers, first in 2015 and then following up with the same farmers in 2016. All farmers acquired their drip-irrigation systems through the government-subsidized NMMI programme over the past few years. As detailed below, the implementation of the NMMI programme in

Rajasthan has been evolving, particularly with respect to subsidy levels. The Government of India (2014) recently evaluated the implementation of the NMMI across a sample of 10 states. While detailing the conclusions of this evaluation is beyond the scope of the present article, the report did conclude that the NMMI application process for farmers in Rajasthan was 'cumbersome' compared to other states and that some of the supplied irrigation equipment was of poor quality. The report recommended that the government of Rajasthan revise its application process to increase the adoption of drip irrigation. While reliable data are not yet available on the total area now under drip irrigation in Rajasthan, one GOI presentation suggested that it has nearly doubled, from 13,104 ha in 2010 to 25,669 ha in 2015 (GOI, 2016).

The findings of the field research presented here indicate three main tensions in the state's desire to reduce groundwater usage in irrigation and enhance climate resilience via the spread of efficiency-enhancing technologies. (1) Farmers primarily adopt drip irrigation to intensify and expand the area under production, to increase production and income. (2) In doing so, they are fully aware that their water demands are growing. with negative impacts on groundwater recharge, despite the state's goal of saving water. (3) All the farmers using drip irrigation in the area were commercial farmers who were able to afford drip irrigation's high initial investment cost and to navigate the cumbersome and sometimes usurious application process for state subsidy. The better ability of commercial farmers to complete the application process and so to benefit from the drip-irrigation subsidy is likely to worsen social-economic disparities between the adopters (who are already relatively well-off) and the non-adopters. Furthermore, these farmers believe that by adopting drip irrigation now and expanding production, they will accrue economic gains that will elude late-adopting farmers. These latter aspects are returned to in the discussion.

This research was conducted in Bassi Tehsil (administrative block) in Jaipur District, about 40 kilometres east of Rajasthan's capital city of Jaipur, towards the city of Agra. Farmers were selected based on a purposeful, snowball sampling technique with the criterion of having adopted a drip-irrigation system. Drip irrigation is still relatively new to the area, and so 12 farmers represented, at the time, nearly the entire population of adopters. All 12 farmers benefited from the drip-irrigation subsidy, which ranged from 70% to 90% of the total cost of the 'drip irrigation system package'. The state government also offers financing directly to farmers to cover their share of the cost, but none applied for or used this financing. This subsidy rate is different from the stated ratio of 50:10:40 as set out in the NMMI because the NMMI grants individual states the leeway to boost subsidy rates to incentivize farmer adoption. Farmers who adopted drip irrigation in 2014 did so with a 90% total subsidy, because the government of Rajasthan was keen to support its diffusion. But in January 2015, the subsidy had declined to 70% after the state government's switch from the Indian National Congress to the Bharatiya Janata Party. This induced farmers who had not yet adopted to wait for the subsidy to rebound back to 90%. The total average cost of the drip irrigation system package for the 12 farmers interviewed in Bassi Tehsil was approximately US$ 2,300 (1.5 lakh rupees) per hectare, after the 90% subsidy.

All the farmers interviewed had adopted drip irrigation with the express purpose of enhancing production through intensification and expansion of their irrigated area. Drip irrigation enabled these farmers to intensify production both by increasing production per unit area and by shortening fallow times (here fallow refers to land left unseeded for at least one growing season). The former is accomplished by reducing the spacing between rows and delivering fertilized irrigation water (fertigation) directly to individual plants. The latter is accomplished by constructing greenhouses over existing fields to enable the production of off-season crops (see below). Further, all 12 farmers interviewed reported no reductions in groundwater pumping time or schedules. Reducing groundwater extraction is not their goal.

One farmer's perspective is illustrative of a general sentiment: 'We went in for the drip to raise productivity. On average, we are getting double the production from previous [flood irrigation methods]' (personal interview with landowner, February 2015). This household in February was growing water-intensive green chilies destined for commercial markets (Figure 1). This raises a number of issues. Clearly, cultivating a water-intensive crop draws into question whether the diffusion of drip irrigation will lead to water savings or, instead, to various forms of intensification, as is the case here. Drip irrigation enables productivity gains via 'more crop per drop', but does not save water, because farmers are not incentivized (or disincentivized) to do so. Another farmer said the next year, 'Formerly, most chilies were grown on the Jodhpur side [western Rajasthan], and they made a lot of money. But now their water is finished. It's gone hard [which is not good for chili production]. Their bore-wells are at least a thousand feet deep. This makes it too costly [to pump up to grow chilies]. So now we are doing this' (personal interview with landowner, July 2016).

Figure 1. Map of Rajasthan, India, with study area shaded.

The reduction of water-demanding chili production in western Rajasthan created a highly remunerative market opening for chili production in Bassi Tehsil. But due to their water-intensive character, their continued production is likely to reduce groundwater level and quality. And this is already the case, according to one of the 12 farmers interviewed: 'We grew chilies some time back, but the [ground]water became too little, and it was hard. So we stopped and switched to other crops that are less sensitive [to irrigation quality or quantity], such as millet in the summer and wheat or cumin in the winter' (personal communication, March 2015). As others have shown with the adoption of new agricultural technologies, the adoption and diffusion of drip-irrigation technology initially allows the production of new kinds of financially remunerative crops, but the very ecological conditions necessary for this expanded production (groundwater quality in this case) are being undermined through this production (Birkenholtz, 2009).

Farmers are also increasing production through a second form of intensification. They are shortening fallow periods through the construction of greenhouses. Greenhouses allow farmers to essentially gain an extra cropping season in the winter months, going from two crops per year to three. The greenhouses protect crops from the cold, and especially from the occasional frost. Farmers grow a combination of summer vegetables and early-summer tomatoes. For instance, they cultivate okra, which is a summer crop, in the winter. These greenhouses allow farmers to grow off-season crops in an effort to benefit from the higher market prices they fetch during the off season. In both cases – new chili production or greenhouse utilization – farmers are using drip irrigation to produce new kinds of water-demanding crops that fill market niches. On the one hand, this signifies their flexibility to meet changing market conditions and demonstrates acumen. On the other hand, it increases their use of water.

Farmers also extend the area under production by bringing more land under the plough and increasing overall irrigated area. Of the 12 farmers interviewed, landholdings ranged from 1.38 to 6.33 ha, with an average of 3.7 ha. Before adopting drip irrigation, the farmers irrigated, on average, nearly 40% of their cultivatable area. After adopting drip irrigation, this increased to 67% on average. According to the farmers, 'the major limiting factor of further expansion [of irrigated area] is the lack of water. Drip irrigation allows us to expand [area under production] because we can irrigate more.... But we are using about the same amount of water [as we did with conventional irrigation] ... or maybe more.' It is reasonable to predict that the irrigated area will continue to increase thanks to drip irrigation. This is likely to increase production but will do little to stem groundwater overdraft.

As further evidence of the desire to increase or expand production, 4 out of the 12 farmers interviewed had also participated in a separate state programme that subsidized the adoption of solar-powered tube-wells. The wells are meant to reduce energy demand in irrigated agriculture, but farmers are using them to increase productivity, in a context where they receive electricity for only eight hours per day, on a day/night alternating basis (Figure 2). Explained one farmer, 'One week we get eight hours of electricity a night, and then the next we get eight hours during the day. When the electricity is at night we run both [the] solar well [during the day] and the regular bore-well [at night]' (personal interview with landowner, July 2016). This is done with the full knowledge that these actions are exacerbating groundwater

Figure 2. Farmer-constructed greenhouse (behind solar panels) to enable off-season crop production.

decline. A second farmer said, 'Water will be depleted [beyond the solar pump] in three years' time.... We will go deeper, or sell' (personal interview with landowner, March 2015). These farmers are knowingly mining groundwater because of a regulatory institutional vacuum in which they are incentivized to extract groundwater both to enhance productivity and to extract groundwater in the near term, before others can. They are also incentivized to continue to raise productivity through the multiple methods of intensification identified above. The motivations of farmers and the conditions under which they elect to adopt drip-irrigation systems are not being considered by state planners. Ultimately, this is leading to continued or even expanded groundwater extraction, rather than water savings. It is to these tensions that we turn in the next section.

Discussion: the rebound effect, efficiency and the political economy of adoption

The broad goals of the Indian state (along with development donor agencies) are to facilitate the adoption and diffusion of drip-irrigation technology to reduce demand for groundwater by enhancing the efficiency of existing irrigation systems. Climate change and rapid groundwater decline serve as the underlying physical conditions and context within which this is occurring. But it also must be remembered that these goals are to be realized largely in the absence of new rules or laws around groundwater abstraction. There are two broad tensions internal to these goals that are undermining their realization. These have to do, first, with the tensions between efficiency and water savings, and, second, between the rationale for farmers to use drip irrigation and the lack of meaningful groundwater regulations.

Overall, the findings presented here demonstrate the operation of the Jevons paradox, or rebound effect, where the adoption of a more technically efficient (in terms of physical productivity or technical efficiency) natural resource technology, in this case drip irrigation, leads to an increase in the use of that natural resource rather than a decrease. One might ask: why wouldn't it? Even though the lion's share of groundwater is drawn for irrigation, the small quantity of water available for irrigation has been an impediment to the further expansion of irrigated area in India. Therefore, when presented with an opportunity to reduce water consumption, while maintaining or even enhancing existing production, farmers see the water savings as a resource that can be reallocated by intensifying cultivation and extending the area under cultivation. Many farm-level studies of drip-irrigation adoption focus on how to promote the further diffusion of drip irrigation through specific policies that will incentivize their adoption, while also enhancing economic and technical efficiency (Friedlander, Tal, & Lazarovitch, 2013; Kulecho & Weatherhead, 2005; Namara, Nagar, & Upadhyay, 2007). Yet, as Venot (2016) clearly demonstrates, the discourse of drip irrigation as a promising climate adaptation technology with the ability to create productive and economic efficiency, while reducing water demand, has become a truism, but needs to be more critically analyzed. As in previous historical moments in the rapid spread of new agricultural technological innovations (e.g., the Green Revolution), drip-irrigation systems are ushering in a new production regime characterized by rising productivity, but at the continued cost of groundwater overdraft.

Second, the continued focus on the technology itself as central to realizing efficiency gains, as well lowering demand for water in irrigation, is illustrated in the tensions between government policies that further reflect the political economy of irrigated agriculture in India. One of the central goals for the National Mission on Sustainable Agriculture (NMSA) is to reduce the water withdrawals for irrigation from more than 90% in India (Shah, 2009) to the 'global norm' of 70% by making irrigation more efficient. Of particular concern for the NMSA is to do this while enhancing food production and increasing the resilience of agriculture to future climate change–related shocks. As I noted above, the central vehicle to accomplish these goals is the smallholder adoption of micro (drip) irrigation systems. This goal is of such importance and centrality to the NMSA that the NMMI was created to realize this express goal, by expanding India's total drip-irrigated area from 2 to 69 million hectares. But the NMMI does not necessarily share the goals of the NMSA; it is focused on expanding the area of irrigation under drip, obscuring the goal of reducing water demand. In this way, neither the NMMI nor adopting farmers explicitly promote or adopt drip irrigation for its ability to reduce demand for water but for its ability to enhance production in a political economic agricultural context where smallholders are attempting to improve their material well-being by increasing agricultural surpluses.

The interaction of these processes, then, has implications for crafting policy that minimizes the rebound effect, while maintaining and even enhancing productive efficiency, in a context of groundwater overdraft and climate change.

Conclusions: policy implications

Drip irrigation is being promoted throughout the world as a climate-adaptive technology that will reduce demand for groundwater and notably enhance the resilience of agriculture in light of ongoing groundwater overdraft and climate change–induced shocks. This article has examined the recent adoption of drip irrigation by farmers in one administrative district of Rajasthan, India, that is undergoing an expansion of drip irrigation. In doing so, it offered a further demonstration of the operation of the Jevons paradox with respect to drip irrigation systems, where the political economy of groundwater-led agriculture incentivizes intensification, which rather than reducing the demand for groundwater is exacerbating groundwater over-extraction.

These findings have a number of policy lessons and implications. First, we need more research focused on understanding what drives the Jevons paradox and what institutional levers may be employed to close the gap between water-use efficiency and intensification dynamics. As with the state's inability to pass groundwater regulation, promoting the adoption of drip-irrigation technology in the absence of any rules governing groundwater use is unlikely to lead to water savings. In the short term, it will lead to enhanced productivity, but in the long term it will lead to further groundwater mining and ultimately to productivity losses. So, too, so-called common sense policy levers, such as increasing the cost of electricity (Scott & Shah, 2004; Shah et al., 2004), are unlikely to reduce groundwater pumping unless farmers are prevented from passing these additional costs to the consumer, thereby reducing their profit margins.

Finally, the Indian National Mission for Sustainable Agriculture, which is the main central-government policy aimed at reducing agricultural demand for groundwater, has two notable flaws. First, it encourages drip and micro irrigation without any consideration of ecological or social resilience in any meaningful way. Here one is supposed to have faith that the diffusion of drip irrigation across the landscape will lead to positive socio-ecological outcomes. Second, there is no mention of the social consequences of the socially uneven adoption of drip irrigation. This has particular importance both for the realization of the development of agrarian livelihoods and for continued agricultural productivity.

Disclosure statement

No potential conflict of interest was reported by the author.

References

Aeschbach-Hertig, W., & Gleeson, T. (2012). Regional strategies for the accelerating global problem of groundwater depletion. *Nature Geoscience*, *5*, 853–861. doi:10.1038/ngeo1617

Benouniche, M., Kuper, M., Hammani, A., & Boesveld, H. (2014). Making the user visible: Analysing irrigation practices and farmers' logic to explain actual drip irrigation performance. *Irrigation Science*, *32*(6), 405–420. doi:10.1007/s00271-014-0438-0

Berbel, J., & Mateos, L. (2014). Does investment in irrigation technology necessarily generate rebound effects? A simulation analysis based on an agro-economic model. *Agricultural Systems*, *128*, 25–34. doi:10.1016/j.agsy.2014.04.002

Birkenholtz, T. (2009). Irrigated landscapes, produced scarcity, and adaptive social institutions in Rajasthan, India. *Annals of the Association of American Geographers*, *99*(1), 118–137. doi:10.1080/00045600802459093

Birkenholtz, T. (2014). Recentralizing groundwater governmentality: Rendering groundwater and its users visible and governable. *Wiley Interdisciplinary Reviews: Water*, *2*(1), 21–30.

Boelens, R., & Vos, J. (2012). The danger of naturalizing water policy concepts: Water productivity and efficiency discourses from field irrigation to virtual water trade. *Agricultural Water Management*, *108*, 16–26. doi:10.1016/j.agwat.2011.06.013

Cullet, P. (2012). The groundwater model bill-rethinking regulation for the primary source of water. *Economic and Political Weekly*, *47*(5), 40–47.

Cullet, P. (2014). Groundwater law in India towards a framework ensuring equitable access and aquifer protection. *Journal of Environmental Law*, *26*(1), 55–81.

Dumont, A., Mayor, B., & López-Gunn, E. (2013). Is the rebound effect or jevons paradox a useful concept for better management of water resources? Insights from the irrigation modernisation process in Spain. *Aquatic Procedia*, *1*, 64–76. doi:10.1016/j.aqpro.2013.07.006

Fishman, R., Devineni, N., & Raman, S. (2015). Can improved agricultural water use efficiency save India's groundwater. *Environmental Research Letters*, *10*, 084022. doi:10.1088/1748-9326/10/8/084022

Friedlander, L., Tal, A., & Lazarovitch, N. (2013). Technical considerations affecting adoption of drip irrigation in sub-Saharan Africa. *Agricultural Water Management*, *126*, 125–132. doi:10.1016/j.agwat.2013.04.014

Gosain, A. K., Rao, S., & Basuray, D. (2006). Climate change impact assessmenet on hydrology of Indian river basins. *Current Science*, *90*(3), 346–353.

Government of India (GOI). (2010). *National mission on micro irrigation - Operational guidelines*. New Delhi: Ministry of Agriculture (Department of Agriculture and Cooperation).

Government of India (GOI). (2014). *Evaluation study on integrated scheme of micro irrigation*. New Delhi: Planning Commission (Programme Evaluation Organisation).

Government of India (GOI). (2016). *Rajasthan agriculture: Road map*. Paper presented at the Meeting of NITI Ayog, Ahmedabad, India. Retrieved from http://niti.gov.in/writereaddata/files/Rajasthan_Presentation_0.pdf

Government of Rajasthan (GOR). (2010). *State water policy - 2010*. Jaipur: Government of Rajasthan.

Government of Rajasthan (GOR). (2011). *Statistical abstract - Rajasthan 2011. Government of Rajasthan: Directorate of economics and statistics*. Jaipur: Government of Rajasthan.

Hooda, S. M. (2014). *Rajasthan water assessment: Potential for private sector interventions. IFC World Bank Group*. New Delhi. Retrieved from http://www.ifc.org/

IGRAC. (2010). Global Groundwater Information System (GGIS). Retrieved from http://www.igrac.net/

IWMI. (2006). *Promoting micro-irrigation technologies that reduce poverty*. Colombo, Sri Lanka. Retrieved from http://www.iwmi.cgiar.org/Publications/Water_Policy_Briefs/PDF/WPB23.pdf?galog=no&iwmi=1

Johansson, R. C., Tsur, Y., Roe, T. L., & Doukkali, R. (2002). Pricing irrigation water: A review of theory and practice. *Water Policy*, *4*, 173–199. doi:10.1016/S1366-7017(02)00026-0

Kulecho, I., & Weatherhead, E. (2005). Reasons for smallholder farmers discontinuing with low-cost micro-irrigation: A case study from Kenya. *Irrigation and Drainage Systems*, *19*(2), 179–188. doi:10.1007/s10795-005-4419-6

Kuppannan, P., Mohan, K., Kakumanu, K. R., & Raman, S. (2011). Spread and economics of micro-irrigation in India: Evidence from nine states. *Economic and Political Weekly*, *46*(26–27), 81–86.

Lankford, B. (2006). Localising irrigation efficiency. *Irrigation and Drainage*, *55*(4), 345–362. doi:10.1002/(ISSN)1531-0361

Lankford, B. (2012). Fictions, fractions, factorials and fractures; on the framing of irrigation efficiency. *Agricultural Water Management*, *108*, 27–38. doi:10.1016/j.agwat.2011.08.010

Li, T. M. (2007). *The will to improve: Governmentality, development, and the practice of politics*. Durham: Duke University Press.

Molle, F., Venot, J.-P., & Hassan, Y. (2008). Irrigation in the Jordan Valley: Are water pricing policies overly optimistic? *Agricultural Water Management*, *95*(4), 427–438. doi:10.1016/j.agwat.2007.11.005

Namara, R., Nagar, R., & Upadhyay, B. (2007). Economics, adoption determinants, and impacts of micro-irrigation technologies: Empirical results from India. *Irrigation Science*, *25*, 283–297. doi:10.1007/s00271-007-0065-0

Narayanamoorthy, A. (2004). Drip irrigation in India: Can it solve water scarcity. *Water Policy*, *6* (2), 117–130.

Narayanamoorthy, A. (2010). Can drip method of irrigation be used to achieve the macro objectives of conservation agriculture? *Indian Journal of Agricultural Economics*, *65*(3), 428–438.

National Geographic. (2013). Drip irrigation expanding worldwide. Retrieved from http://newswatch.nationalgeographic.com/2012/06/25/drip-irrigation-expanding-worldwide/

Parsai, G. (2013, May 13). Centre to establish National Bureau of water use efficiency. *The Hindu*.

Pei, H. W., Scanlon, B. R., Shen, Y. J., Reedy, R. C., Long, D., & Liu, C. M. (2015). Impacts of varying agricultural intensification on crop yield and groundwater resources: Comparison of the North China Plain and US High Plains. *Environmental Research Letters*, *10*(4), 044013. doi:10.1088/1748-9326/10/4/044013

Perry, C. (2007). Efficient irrigation; inefficient communication; flawed recommendations. *Irrigation and Drainage*, *56*(4), 367–378. doi:10.1002/(ISSN)1531-0361

Pullabhotla, H. K., Kumar, C., & Verma, S. (2012). Micro-irrigation subsidies in Gujarat and Andhra Pradesh [India] implications for market dynamics and growth. Retrieved from http://www.iwmi.cgiar.org/iwmi-tata/PDFs/2012_Highlight-43.pdf?galog=no&iwmi=1

Rose, N. (1999). *Powers of freedom: Reframing political thought*. Cambridge: Cambridge University Press.

Scott, C. A., & Shah, T. (2004). Groundwater overdraft reduction through agricultural energy policy: Insights from India and Mexico. *International Journal of Water Resources Development*, *20*(2), 149–164. doi:10.1080/0790062042000206156

Shah, T. (2009). Climate change and groundwater: India's opportunities for mitigation and adaptation. *Environmental Research Letters*, *4*(3), 035005. doi:10.1088/1748-9326/4/3/035005

Shah, T., Scott, C., Kishore, A., & Sharma, A. (2004). *Energy-irrigation nexus in South Asia: Improving groundwater conservation and power sector viability* (Vol. 70). Colombo: International Water Management Institute.

Siebert, S., Burke, J., Faures, J. M., Frenken, K., Hoogeveen, J., Doll, P., & Portmann, F. T. (2010). Groundwater use for irrigation - a global inventory. *Hydrology and Earth System Sciences*, *14* (10), 1863–1880. doi:10.5194/hess-14-1863-2010

Taylor, R. G., Scanlon, B., Döll, P., Rodell, M., van Beek, R., Wada, Y., … Treidel, H. (2013). Ground water and climate change. *Nature Climate Change*, *3*(4), 322–329. doi:10.1038/nclimate1744

Tiwari, V. M., Wahr, J., & Swenson, S. (2009). Dwindling groundwater resources in northern India, from satellite gravity observations. *Geophysical Research Letters*, *36*, L18401. doi:10.1029/2009gl039401

van der Kooij, S., Zwarteveen, M., Boesveld, H., & Kuper, M. (2013). The efficiency of drip irrigation unpacked. *Agricultural Water Management*, *123*, 103–110. doi:10.1016/j.agwat.2013.03.014

Venot, J.-P. (2016). A success of some sort: Social enterprises and drip irrigation in the developing world. *World Development*, *79*, 69–81. doi:10.1016/j.worlddev.2015.11.002

Ward, F. A., & Pulido-Velazquez, M. (2008). Water conservation in irrigation can increase water use. *Proceedings of the National Academy of Sciences*, *105*(47), 18215–18220. doi:10.1073/pnas.0805554105

World Bank. (2014). Improving Punjab irrigation: More crops from every drop. Retrieved from http://www.worldbank.org/en/news/feature/2014/04/18/improving-punjab-irrigation-more-crops-from-every-drop

Worldwatch Institute. (2013). "Efficient" irrigation tool may deplete more water. Retrieved from http://www.worldwatch.org/node/5942

Yadav, R. K., Kumar, K. R., & Rajeevan, M. (2010). Climate change scenarios for Northwest India winter season. *Quaternary International*, *213*(1–2), 12–19. doi:10.1016/j.quaint.2008.09.012

Climate change, groundwater and the law: exploring the connections in South Africa

Michael Kidd

ABSTRACT
Projected impacts of climate change on water availability in South Africa are likely to result in the increasing use of groundwater, which is relatively underused at present. Several threats to groundwater, including acid mine drainage, pervasive water pollution (particularly from untreated sewage), and planned hydraulic fracturing will have to be addressed to protect the country's groundwater reserves. This article considers the role that law can play in both managing groundwater and protecting it from these and other threats.

Introduction

South Africa is already a water-scarce country (DWAF, 2004), and it is projected that climate change will exacerbate the situation. Most of the water used in the country currently is surface water (DWA, 2010a), which is probably most at risk from climate change impacts. There is thus considerable potential for increased reliance on groundwater for productive use in the country, if it can be protected from its considerable threats (including acid mine drainage, eutrophication and pollution by untreated sewage, and potentially hydraulic fracturing). This highlights the need for good groundwater governance, which can be defined as 'the process by which groundwater is managed through the application of responsibility, participation, information availability, transparency, custom, and rule of law. It is the art of coordinating administrative actions and decision making between and among different jurisdictional levels' (Varady et al., 2012, p. 7).

This article will consider the legal aspect of groundwater governance – in other words, the role that law can play in South Africa in conserving groundwater and ensuring that such water can be used to address both current needs and those that will be presented by climate change impacts into the future. While the current water law in South Africa applies equally to both surface and groundwater, the role of the law in the management and allocation of groundwater resources depends on resource-intensive data collection and evaluation, which has proved to be an onerous and protracted process in the experience of the implementation of South Africa's water law thus far. Although there are legal mechanisms available for addressing threats to groundwater, there are implementation challenges, both actual and potential. Even if well implemented, however, the law can at best play a supporting role in addressing some of the main threats to the integrity of South Africa's groundwater resources, such as acid mine drainage, sewage contamination and planned hydraulic fracturing. Other governance tools beyond the scope of this article, such as research, gathering of relevant information on groundwater resources, and provision of finance will have to play their part as well.

South Africa's water situation

The National Water Resources Strategy 2013 (DWA, 2013) provides the following facts about South Africa's water context. Overall, the country is the 30th driest country in the world, but not uniformly so: 'The variable rainfall distribution and characteristics give rise to the uneven runoff and distribution of water resources across the country, with more than 60% of the river flow arising from only 20% of the land area.' Many catchments in the country are already approaching (or have already reached) a situation where supply cannot meet demand (see also Kidd, 2011). Available surface water (more than 80% of which has already been allocated) is between 9500 million m^3/y and 12,000 million m^3/y, about a third of which is moved via interbasin transfers, and there are 4395 registered dams in the country. Available renewable groundwater, on the other hand, is between 7500 million m^3/y (high assurance, meaning that which is available under drought conditions) and 10,343 million m^3/y. Compared to the amount available, current groundwater use is about 2000 million m^3/y, which means that 'allowing for an underestimation on groundwater use, about 3500 million m^3/a could be available for further development' (DWA, 2013). The eastern parts of the country are wetter than the interior and the west, much of the latter being arid. Groundwater availability largely mirrors the surface water situation (i.e. the west has less of both surface and groundwater).

This was the situation in the recent past, which is exacerbated today by South Africa's worst drought in 30 years (Essa, 2015). Water availability will become even more critical in the near future, as both climate change and other developments such as increasing population and planned economic growth add pressures to existing water reserves.

As for groundwater in particular, according to the National Water Resources Strategy (DWA, 2013, p. 25),

> Groundwater is the primary source of reliable, safe drinking water supplies in rural areas and for many towns in South Africa; for the irrigation of thousands of hectares of valuable arable land around the country; and for supporting large numbers of livestock and game. Many mines and industries also rely on groundwater for their supplies.

More specifically, King, Maree, and Muir (2009) state that more than 280 towns and settlements in the country are largely dependent on groundwater, and that it has been estimated that '78% of all groundwater abstracted is used for irrigation, 7% for rural domestic purposes and 6% for stock watering. Only 4% of groundwater abstracted is used in the urban environment' (p. 435).

South Africa's water imperatives

Though most of the country's surface water has already been allocated, the population is increasing, and planned economic growth and social development will increase the demand for water (DWA, 2013). In urban areas, only 37% of towns are assessed as having no water shortage for 10 years, whilst 25% are projected as having a shortage at some stage over the next 10 years. Some 30% are currently in deficit (DWA, 2013, p. 23). Groundwater is seen as an important component of the overall strategy aiming at reconciling demand and supply in the urban sector (DWA, 2013).

Agriculture (irrigation) is currently the biggest user of water in South Africa (DWA, 2013), and the National Development Plan 2013 (National Planning Commission, 2011) stresses that key proposals in the agriculture and agro-processing sectors include a considerable increase in irrigation infrastructure. It envisages an increase by a further 500,000 hectares of the land under irrigation by 2030 from the current 1.5 million hectares. This will be achieved through a combination of 'better use of existing water resources and developing new water schemes' (p. 219). Groundwater will thus clearly have an important role to play in providing water, not only for urban needs, but also for increased irrigation.

Water and climate change in South Africa

South Africa's *White Paper on the National Climate Change Response* (DEA, 2011, p. 9) indicates that 'parts of the country will be much drier and increased evaporation will ensure an overall decrease in water availability'. Bearing in mind that southern Africa is an area where rainfall is already highly variable, climate change projections for this region 'suggest that dry areas will become even drier and rainfall more erratic' (Milgroom & Giller, 2013, p. 91). According to Kusangaya, Warburton, van Garderen, and Jewitt (2014, p. 48), several scholars have indicated that there is 'certainty' that climate change will 'impact on the availability and use of water resources'. The same authors (p. 50) state that the 'general conclusion' from most studies in the region is that streamflow will decrease by 2050, although localized effects may well differ, ranging from projected decreases of up to 75% (in the Pungwe catchment – Andersson, Samuelsson, & Kjellstro, 2010) to projected increases of 16–38% (in the Thukela catchment – Graham et al., 2011).

The white paper suggests that the drier western and interior parts of the country are likely to become more arid, whilst the comparatively wetter eastern parts of the country are likely to become more humid, with increased instances of floods and droughts. Surface water will become prone to increased rates of evaporation as it becomes hotter, which will have a negative impact on surface storage.

In addressing key adaptation issues, the white paper indicates that water reconciliation (i.e. balancing supply and demand), an already difficult task (as outlined above), will become even more so with the effects of climate change). It identifies key elements of climate change response (adaptation) for the water sector, including improvement of data relating to water, water conservation and water demand management, and 'exploring new and unused resources, particularly groundwater, re-use of effluent, and desalination' (p. 17). The 2013 National Water Resources Strategy echoes this in stating that the DWA 'will consider all appropriate sources of water for increasing water supply, including groundwater and alternative water supply sources. The use of these will be tested against the climate change scenarios' (DWA, 2013, p. 78).

Similar ideas are reflected in the 2010 *Groundwater Strategy* (DWA, 2010a, p. 8), where the importance of groundwater in addressing the impacts of climate change is highlighted:

> The advantages of groundwater (much lower evaporation/evapotranspiration and slower declines in drought years, because the volumes stored underground are so much higher compared with surface water) mean that it should form a key part of our strategy to adapt to climate change.

It is not as simple as drilling boreholes, however, as there is a substantial amount of scientific information relating to both quantity and quality of groundwater that has to be compiled. For example, there is still a lot of work that 'needs to be done regarding the effects of climate change on technical issues such as ground water recharge' (DWA, 2010a, p. 8).

The above discussion of the context within which the law operates in South Africa indicates, broadly, that the current shortages experienced in many parts of the country will be exacerbated by climate change, and this will require increased use of groundwater in the country. How is South Africa's water law (which is the focus of this article, although other laws are of potential relevance) equipped to facilitate this in such a way that the available groundwater reserves are used sustainably?

Relevant South African law

Prior to colonial legislation, water use was regulated by customary law, which is still observed in many parts of South Africa (Kapfudzawura & Sowman, 2009). Customary law does fall within the definition of 'existing lawful use' in the National Water Act (discussed further below), although the content of these rules focused primarily on disputes (Tewari, 2009) and their direct applicability to groundwater is not clear.

The Water Act 54 of 1956, which was the water legislation in force at the time of South Africa's transition to democracy in 1996, was originally enacted to provide for increased government control over water resources, but these controls were rarely used, with the effect that most users continued using water on the basis of common law (riparian) rights. As De Wet observed in 1959 (p. 35), the act was a 'half-hearted attempt to restore to the community the rights lost by a process of judicial legislation, and the doctrine of "riparian rights" is by no means dead'. The effect of this is that the government had very little control over the use of water by riparian users (even though control mechanisms were available) and that many groups of the population had limited, if any, access to water.

The act distinguished between public water and private water, the latter essentially requiring no balancing of use rights with other users. It defined 'private water' as all water that rises or falls naturally on any land or naturally drains or is led onto one or more pieces of land, but is not capable of common use for irrigation purposes (§ 1). The owner of the land on which private water was found had the exclusive use of the water (§ 5(1)), but intentional or unintentional pollution was prohibited (§ 23). The sale or disposal of private water was prohibited, except under authority of a permit from the Minister of Water Affairs (§ 5(2)).

The act defined 'public water' as any water flowing or found in or derived from the bed of a public stream, whether visible or not (§ 1). A 'public stream' was defined (in essence) as a natural stream of water that flows in a known and defined channel, even if dry during part of the year, if the water is capable of common use for irrigation on two or more pieces of riparian land (§ 1). The right to use public water for agricultural and urban purposes (terms defined in the act) was vested in the riparian owner. This owner had a share of the normal flow of the water in the public stream, as fixed by the water court (§§ 9(10), 52). A riparian owner was authorized to use all the surplus water – water in a public stream that is not normal flow (§ 1) – for beneficial agricultural or urban purposes and was not required to abate such use in favour of other riparian owners (§ 10(1)). A person could use public water for defined purposes only to the extent that the use was beneficial (§ 10(1)). Wasteful use was prohibited (§§ 9(1)(a), 170). There was no fee for using water extracted directly from a public or private water source.

Groundwater could be either public or private if it fitted within the relevant definition, in which case the act governed it. Groundwater could also be 'subterranean water' if found in a subterranean government water control area and hence under state control, or it could be that it was neither public, nor private water, nor under state control (Lyster & Lazarus, 1995), in which case it was subject to common-law principles. There were a handful of subterranean government water control areas, so at the commencement of the 1998 National Water Act (see below) most groundwater was not within one of these areas (DWA, 2010a). In other words, in 1999 (when the new act commenced) most groundwater was not under state control and was accessed largely on the basis of the common-law approach of riparianism.

An examination of the pre-1996 law is not only relevant from a historical perspective, but is also important because it is still recognized in relation to certain water uses (as discussed below). In 1996, the new National Water Act 36 of 1998 (NWA) was enacted with two primary aims: sustainable management of water and more equitable access to water resources. The latter objective was required by the fact that the vast majority of access rights to productive water resources in 1996 were in white hands and that millions of South Africans (primarily black) had insufficient access to water (DWAF, 1997). Because it was considered unfeasible for the legal system of access to water resources to change overnight from an overwhelmingly riparian-rights approach to the favoured new water use licensing system, existing lawful use (essentially lawful use for a period of two years prior to the commencement of the 1998 act) was recognized in the new act as a use that would not require a licence, at least initially (Van Koppen & Schreiner, 2014). There has as yet been little progress in transforming existing lawful use to statutorily licensed water use (Van Koppen & Schreiner, 2014), with the result that many users of groundwater (and surface water) in 2015 are still using such water as a result of what was in 1999 (when the NWA came into operation) an existing lawful use right.

Under the 1996 Constitution of the Republic of South Africa, water management is vested in the national government, through the department responsible for water affairs (which is, at the time of writing, called the Department of Water and Sanitation). Only the national legislature has the power to legislate on water matters. When it comes to administration and implementation of several aspects of water management, however, local government also has an important executive role, in relation to provision of water services (Part B of Schedule 4 of the Constitution provides that 'water and sanitation services limited to potable water supply systems and domestic waste-water and sewage disposal systems' are a local government executive task). For purposes of this article, however, the relevant decision-making powers are those of the national Department, in some cases operating through provincial branches of the national entity. In relation to the water-allocation process, municipalities are allocated water to meet their constitutional obligations as well as obligations in terms of various acts of parliament, such as the Water Services Act 108 of 1997. Municipalities do not make allocation decisions (but, like any user, they have the right to participate in the decision-making process through procedural fairness procedures).

The NWA provides for the concept of water use in § 21 – not just abstraction of water but also any act that has an impact on water, including diverting watercourses and disposing of effluent into water. Any water use requires a licence from the national department, unless it is a Schedule 1 use (explained below), an existing lawful use or a general authorization (which is essentially a blanket licence for stipulated uses). Schedule 1 uses are basically *de minimis* water uses, including domestic use and watering of animals that are not part of intensive farming. Because of this legal regime, many users of groundwater in South Africa in 2016 are using water legally but without a licence (for domestic use, for example).

The intention behind the legal regime introduced by the NWA was that existing lawful uses would be replaced in due course by licensed uses, taking into account the elaborate considerations envisaged by the act (Van Koppen & Schreiner, 2014). This process has been slow, however, with not much progress made since 1998. The act provides a process of compulsory licensing (§ 43) to replace existing lawful use, but at the time of writing there have been only three such processes, covering a small geographical area in relation to the overall area of the country, and a small water-volume portion of overall water usage. One such process, the Tosca Molopo process, involved exclusively groundwater and is considered in further detail below.

All licensing (for use of any water – surface or ground) is based on a complex, resource-intensive and time-consuming planning process of determining the classification of the water resource in question (primarily in relation to existing water quality vis-à-vis water resource quality objectives, (§§ 12–15), followed by determination of the 'reserve' (§§ 16–18), which is defined as the quantity and quality of water required to satisfy basic human needs and to protect aquatic ecosystems to secure ecologically sustainable development and use of the relevant water resource (§ 1). Once the reserve has been determined, in addition to the amount of water required for international commitments, the remaining water is the amount that is available for allocation by means of licensing.

For purposes of this article, it is important to note that the NWA treats water as part of a holistic hydrological system, which means that all water, surface and ground, is treated by the law in the same way. Consequently, if a user (including both natural persons and municipalities) wants to use groundwater, and it is not a Schedule 1 (*de minimis*) use, or subject to a general authorization (which does set out conditions for the water use in question), that user will require a licence. The national department decides on licence applications on the basis of the classification of the water resource in question, its reserve determination, and the factors set out in § 27 of the act. These are listed in an open-ended set of considerations which include the need to redress the results of past racial and gender discrimination; the efficient and beneficial use of water in the public interest; and the likely effect on the water resource and on other water users of the water use to be authorized. The requirements of water use licences (§ 28), conditions of issue (§ 29), and review of licences (§ 49) are also regulated by the act.

The law thus requires that new uses of groundwater that are not Schedule 1 uses require licences, and any existing use of groundwater may be subject to a compulsory licensing process, the likely outcome of which will be a reduction in the amount of water currently being used by the user in question. In any licensing process (whether new or as a result of a decision to institute compulsory licensing in an area), the decision to grant a licence, including the determination of the amount of water to allocate and the conditions applicable to the water use, is a decision that rests on the availability of a considerable amount of scientific and other information about the water resource in question (such as the total amount of water available in the area, the number of current users and the amount of water that they are using) and evaluation of that information. In many areas, the process of gathering this information is in varying stages of advancement, and often the information available is inadequate (DWA, 2010a). Pietersen et al. (2012, p. 454) have observed that 'inadequate information on groundwater' is a weakness relating to the necessary knowledge underpinning groundwater governance in South Africa.

Example: the Tosca-Molopo compulsory licensing process

The NWA provides for compulsory licensing in specified geographical areas to achieve a fair allocation of water from a water resource which is under water stress, or when it is necessary to review prevailing water use to achieve equity in allocations; to promote beneficial use of water in the public interest; to facilitate efficient management of the water resource; and/or to protect water resource quality (§ 43). One of three processes that have been carried out in terms of § 43 at the time of writing was applied to the small Tosca Molopo area in the North-West Province, adjoining the northern border of South Africa with Botswana. The major aim of the process was to alleviate water stress, the entire water source in the area being groundwater. The process entails balancing the amount of water that has been requested in the licence applications with the amount of water that is available, taking into account the amount of water that is necessary for the reserve and any relevant international commitments, based on bilateral or multilateral agreements with neighbouring countries sharing the resources (DWAF, 2007). In the case of Tosca Molopo, it was decided that there were no international commitments, and the ecological component of the reserve was considered zero because it was determined that no groundwater contributed to the surface water flow (DWA, 2010b – but this document provides no evidence or other support for this assertion). This, together with the relatively small area and small number of applicants, made it an easier process than most others would be.

The process was finalized without any appeals against the preliminary allocations, which suggests that the process and the objectives behind the process (in this instance, primarily to alleviate water stress) had been communicated clearly to the applicants involved and that the latter had appreciated the need for the process (Seetal, 2012).

Assessment of South African water law relating to groundwater management

The above discussion of the law generally, and of how it was used in the Tosca-Molopo compulsory licensing process, demonstrates that there are legal provisions, on paper, that can be used for the equitable allocation of groundwater resources in a sustainable manner where such resources have not already been allocated, and also in identified cases (compulsory licensing) where they have already been allocated. Given the importance of groundwater in addressing both current and climate change–induced pressures on water resources, the law's regulation (at least on paper) of the allocation of groundwater resources is necessary. Reality, however, is not the same as what appears on paper, and ensuring that decisions are made in accordance with the comprehensive planning regime required by the act is both time-consuming and resource-intensive (see e.g. Movik, 2011). Not only does this require considerable scientific input in relation to aspects such as resource classification and determination of the reserve, but there is a large amount of data that has to be gathered in relation to existing uses that users are often reluctant to divulge accurately (Movik, 2011). This means that the decision-making process that underpins the law's regulation of water, including groundwater, is a protracted process and not necessarily something that will be able to respond quickly to water imperatives in the future. It also, in part, explains why there have been only three compulsory licensing processes in 17 years of the NWA's existence.

In addition to the scientific/technical work outlined above that is preliminary to most water allocation decisions, the situation would be more complex in cases of transboundary groundwater resources. The 2010 *Groundwater Strategy* (DWA, 2010a, p. 25) identifies seven 'important' transboundary aquifers (some of them shared by more than two neighbouring countries). In relation to water resources that are shared by other countries, the NWA is replete with references to South Africa's 'international obligations' (see e.g. §§ 2, 6, 9, 27 and 45), and it is clear in the process of water allocations that international obligations ought to be (at least on paper) regarded as a priority allocation that must be taken into account (where applicable) before domestic allocation decisions may be made.

In relation to groundwater, the international legal situation is not satisfactory. According to Mechlem (2012),

> The most authoritative treaty on international water law, the 1997 United Nations Watercourses Convention ..., formally applies to most shared aquifers but its provisions exclude certain types of aquifers, most importantly non-recharging aquifers. In addition, its substantive provisions are exclusively geared towards surface water and completely ignore the specific management challenges posed by groundwater.

In addition to these shortcomings, only South Africa and Namibia amongst the southern African countries have ratified the convention (UN, 2016). Other applicable international/regional law relating to shared watercourses in South Africa is the Revised Protocol on Shared Watercourses in the Southern African Development Community, which according to Couzens (2016) is underutilized by states in the community, though almost all of them are parties. As far as South Africa and groundwater is concerned, the efficacy of the international legal regime has yet to be put to the test. The significance of these observations is that there is unlikely to be a ready-made template for dealing with sharing of groundwater resources that have an international dimension when allocation decisions have to be made. This is likely to be a basis for further protraction of important decisions in response to climate change impacts, in addition to those pointed out above.

Threats to groundwater

If it were not already difficult enough to license groundwater uses, as outlined above, it is also necessary to take into account several significant threats to groundwater in South Africa, some more concentrated in specific areas of the country, and others pervasive. If these threats to groundwater materialize or if there is a deterioration of the situation in cases where there is already an impact, then the ability of the untapped groundwater potential to address adverse climate impacts in South Africa may be seriously compromised.

Acid mine drainage

Acid mine drainage (AMD) is a major international problem associated with the mining of certain minerals, including gold and coal. It is produced when 'sulfide-bearing material is exposed to oxygen and water' (Akcil & Koldas, 2006), sometimes a natural process but exacerbated by mining, which increases the quantity of sulphides exposed. AMD is a serious problem in South Africa, described as 'one of the greatest environmental threats ever faced by South Africans' (Durand, 2012). AMD, containing high levels of salts, typically having a pH of 2–3 (the pH of pure water is approximately 7, whereas white vinegar has a pH of between 2 and 3), plus heavy metals, and sometimes radioactive, has been decanting to the surface since 2002 (Wells et al., 2009). South Africa's Department of Mineral Resources claims that there are over 8000 derelict and ownerless mines in South Africa, which would take approximately 800 years to rehabilitate, at a cost of billions of rand (Durand, 2012; at the time of writing, approximately 14.5 South African rand equal 1 US dollar). If one adds to this figure 'the rehabilitation of the water bodies affected by AMD, the health impacts on humans, the disruption of societies and the destruction of the biodiversity, the cost becomes incalculable' (Durand, 2012). Also, decanting AMD is a threat to (and has already degraded) surface water. Groundwater is obviously particularly vulnerable, especially in the mining areas (primarily the provinces of Gauteng, North-West and Mpumalanga).

The extent to which the law can be used to combat this problem is unknown. Theoretically, legal provisions can provide measures to alleviate post-mining AMD concerns where mining starts now; but AMD caused by historical mining, much of which was carried out by operators now difficult or impossible to trace or hold accountable, is a problem that the law has a limited role in addressing (Kidd, 2011). More extensive discussion of this problem is beyond the scope of this article, but the threats of AMD to water resources are clearly beyond the ability of law alone to address (Feris & Kotzé, 2014). The huge costs involved, coupled with the difficulty (if not impossibility) of attributing legal liability in relation to historic polluters, means that progress in this regard is slow.

Eutrophication and untreated sewage

Many of South Africa's water resources (rivers and impoundments, primarily) suffer from severe eutrophication (Harding, 2015). One of the main reasons for this is because numerous water-treatment works, most of which are operated by local government bodies, are not carrying out their legal and constitutional responsibilities (Kidd, 2016).

Though the eutrophication is primarily a surface water phenomenon at present, the recharge of groundwater by surface water means that groundwater could be contaminated by the same pollutants that are contributing to problems on the surface. This appears, for example, to be an issue in the Botleng Aquifer in the Delmas area in Gauteng (Pietersen et al., 2012).

From a legal perspective, the water-treatment works mentioned in the previous paragraph are failing to comply with existing water-treatment standards in both the NWA and the Water Services Act 108 of 1997. The prospects of addressing this are very complex (Kidd, 2011), even though the law, on paper (once again), would appear to contain mechanisms to do so. Apart from the political considerations (because the primary offenders are government bodies), the main obstacle to addressing this is financial and not legal (Kidd, 2011). In many cases, sewage pollution is due to the incapacity of many water treatment plants of treating the current levels of waste (Kidd, 2016; Tempelhoff, 2009). The prospect in most cases of sufficient funding to upgrade these plants, given widespread lack of the requisite political will, is not apparent (Kidd, 2016).

Hydraulic fracturing

Many parts of the country, including many of its driest parts (which are the lowest reserves of groundwater), are being earmarked for hydraulic fracturing, or 'fracking' (Van Wyk, 2014). Even though there is still some way to go before the process starts (if it does, although many would say this is a case of 'when', not 'if'), it would be a major threat to groundwater reserves in those parts of the country where it occurs. This is not only because of the threat of contamination of groundwater reserves by the fracking process, but also because huge amounts of water will be required for the operations, some of which may have to be sourced from existing groundwater reserves. It almost goes without saying that the potential impact of hydraulic fracturing on groundwater in drier parts of the country (from both abstraction and possible pollution) for the substantial numbers of (particularly) poor people in these areas will be severe.

From a legal perspective, hydraulic fracturing would appear to be well regulated, at least on paper. First, in terms of Section 38(1) of the National Water Act, the 'exploration and/or production of onshore naturally occurring hydrocarbons that require stimulation, including but not limited to hydraulic fracturing and/or underground gasification, to extract, and any activity incidental thereto that may impact detrimentally on a water resource' has been declared a controlled activity (RSA 2015b). A controlled activity requires a licence under the act. Hydraulic fracturing, from the prospecting to the operational phases, would also require authorization, following an environmental impact assessment process, per Section 24 of the National Environmental Management Act 107 of 1998. The third legal regulatory measure pertaining to hydraulic fracturing is a set of measures aimed at the process which constitute an amendment in 2015 to 'regulations for petroleum exploration and production' made in terms of the Mineral and Petroleum Resources Development Act 28 of 2002 (RSA 2015a). There has not yet been any hydraulic fracturing in the country to test these laws.

At first glance, this would appear to be an appropriate safeguard against undue environmental damage (including adverse impacts on groundwater), but there are concerns about the weight that environmental considerations would carry in the assessment of such projects. Hydraulic fracturing is seen as an economic 'game changer' by the South African government (Anon, 2014; but cf. WWF-SA 2015, which provides some cautionary notes as to its potential economic benefit), and there is a pattern of economic considerations frequently trumping environmental ones in South African environmental authorization processes. Also, the environmental authorization in these cases would be decided by the Minister of Mineral Resources, who is simultaneously tasked with promoting mineral resource development and minimizing the environmental impacts of mining, a difficult balancing act. In view of the numerous approvals of mining operations in important watershed areas in South Africa, the fact that there are legal requirements that have to be met by future hydraulic fracturing operations is somewhat cold comfort (CER 2016).

Overall assessment of legal regulation of threats to groundwater in South Africa

Two of the three main threats to groundwater in South Africa are already present, and the third is likely to arise in the future. In all three cases, there are laws on paper that are applicable to the threatening activities. In relation to AMD and sewage contamination, however, the laws, even if implemented, are insufficient in themselves to address the problems. Both implementation of the law and providing financial resources to address these problems are subject to political will, however, and this this often supports economic development while paying lip service to environmental considerations.

Conclusion

Climate change impacts in South Africa will result in the inevitable increase in the use of groundwater in the not-too-distant future. From a policy perspective, South African authorities would appear to be aware of the likely impact of climate change on water resources in the country, and the role that groundwater may be able to play in alleviating these impacts. The existing legal provisions that are relevant to the use and conservation of groundwater seem to be, at least on paper, capable of addressing sustainable use of groundwater into the future. As is the case in so many countries, however, the implementation gap (the gap between law on paper and law in action) is often alarmingly large in the South African water management context (see also Pietersen et al., 2012). If the law is to operate as intended, there is a need to gather a comprehensive bank of information on groundwater (including extent, quality, and recharge rates) in order for appropriate administrative (i.e. licensing) decisions to be made under the NWA. The law is also required to assist in addressing the threats to groundwater, identified above, although the task is clearly too large just for legal solutions. This is true in relation to both AMD and the problems of untreated sewage, both of which need significant financial resources for adequate response.

Disclosure statement

No potential conflict of interest was reported by the author.

References

Akcil, A., & Koldas, S. (2006). Acid Mine Drainage (AMD): Causes, treatment and case studies. *Journal of Cleaner Production, 14*, 1139–1145. doi:10.1016/j.jclepro.2004.09.006

Andersson, L., Samuelsson, P., & Kjellstro, E. (2010). *Assessment of* climate change impact on water resources in the Pungwe River Basin. S –601 76 Norrköping, Sweden: Meteorological and Hydrological Institute.

Anon. (2014, February 13). Shale gas exploration will be a game changer, says Zuma. *SA News*. Retrieved from http://www.sanews.gov.za/south-africa/shale-gas-exploration-will-be-game-changer-says-zuma

CER (Centre for Environmental Rights). (2016). Zero hour: Poor governance of mining and the violation of environmental rights in Mpumalanga. Retrieved from http://cer.org.za/news/zero-hour

Couzens, E. 2016. *Water-related* conflict and security in Southern Africa: The SADC protocol on shared watercourses (Legal Studies Research Paper). Sydney Law School. Retrieved from http://ssrn.com/abstract=2711496

De Wet, J. C. (1959). One hundred years of water law. *Acta Juridica, 1959*, 31–35.

DEA (Department of Environmental Affairs). (2011). *White* paper on the national climate change response. Pretoria: South African Government.

Durand, J. F. (2012). The impact of gold mining on the witwatersrand on the rivers and Karst system of Gauteng and North West Province, South Africa. *Journal of African Earth Sciences, 68*, 24–43. doi:10.1016/j.jafrearsci.2012.03.013

DWA (Department of Water Affairs). (2010a). *Groundwater* strategy. Pretoria: South African Government.

DWA (Department of Water Affairs). (2010b). *Water* use in the Tosca Molopo catchment: Towards compulsory licensing. Pretoria: South African Government.

DWA (Department of Water Affairs). (2013). *National* water resources strategy (2nd ed.). Pretoria: South African Government.

DWAF (Department of Water Affairs and Forestry). (1997). *White* paper on a national water policy for South Africa. Pretoria: South African Government.

DWAF (Department of Water Affairs and Forestry). (2004). *National* water resource strategy. Pretoria: South African Government.

DWAF (Department of Water Affairs and Forestry). (2007). *A* toolkit for water allocation reform. Pretoria: South African Government.

Essa, A. (2015, November 5). South Africa in midst of 'epic drought'. *Mail & Guardian*. Retrived from http://mg.co.za/article/2015-11-05-south-africa-in-midst-of-epic-drought

Feris, L., & Kotzé, L. J. (2014). The regulation of acid mine drainage in South Africa: Law and governance perspectives. *Potchefstroom Electronic Law Journal, 17*(5), 2105–2163.

Graham, L. P., Andersson, L., Horan, M., Kunz, R., Lumsden, T., Schulze, R. E., … Yang, W. (2011). Using multiple climate projections for assessing hydrological response to climate change in the Thukela River Basin, South Africa. *Physics and Chemistry of the Earth*, Parts A/B/C, *36*, 727–735. doi:10.1016/j.pce.2011.07.084

Harding, W. R. (2015). Living with eutrophication in South Africa: A review of realities and challenges. *Transactions of the Royal Society of South Africa, 70*(2), 155–171. doi:10.1080/0035919X.2015.1014878

Kapfudzawura, F., & Sowman, M. (2009). Is there a role for traditional governance systems in South Africa's new water management regime? *Water SA, 35*(5), 683–692.

Kidd, M. (2011). 'Poisoning the right to water in South Africa: What can the law do? *International* Journal of Rural Law and Policy. doi:10.5130/ijrlp.i1.2011.2604

Kidd, M. (2016). Muddied water: (Un)cooperative governance and water management in South Africa. In J. Gray, C. Holley, & R. Rayfuse (Eds.), *Trans-jurisdictional* water law and governance (pp. 157–174). Abingdon: Earthscan.

King, N. A., Maree, G., & Muir, A. (2009). Freshwater systems. In H. A. Strydom & N. D. King (Eds.), *Fuggle & Rabie's* environmental management in South Africa (2nd ed., pp. 425–454). Cape Town: Juta.

Kusangaya, S., Warburton, M. L., van Garderen, E. A., & Jewitt, G. P. W. (2014). Impacts of climate change on water resources in Southern Africa: A review. *Physics and Chemistry of the Earth*, Parts A/B/C, *67-69*, 47–54. doi:10.1016/j.pce.2013.09.014

Lyster, R., & Lazarus, P. (1995). The problem with ground water in South African law. *South African Law Journal, 112*, 440–461.

Mechlem, K. 2012. *Groundwater* governance: A framework for country action (Thematic Paper 6: Legal and Institutional Frameworks). Rome: Groundwater Governance.

Milgroom, J., & Giller, K. E. (2013). Courting the rain: Rethinking seasonality and adaptation to recurrent drought in semi-arid Southern Africa. *Agricultural Systems, 118*, 91–104. doi:10.1016/j.agsy.2013.03.002

Movik, S. (2011). Allocation discourses: South African water rights reform. *Water Policy, 13*, 161–177. doi:10.2166/wp.2010.216

National Planning Commission. (2011). *National* development plan: Vision for 2030. Pretoria: South African Government.

Pietersen, K., Beekman, H. E., Holland, M., & Adams, S. (2012). Groundwater governance in South Africa: A status assessment. *Water SA, 38*(3), 453–460. doi:10.4314/wsa.v38i3.11

RSA. (Republic of South Africa). (2015a). GN R466 in GG 38855 of 3 June 2015.

RSA. (Republic of South Africa). (2015b). GN 999 in GG 39299 of 16 October 2015.

Seetal, A. (2012, November). Compulsory licensing 'first ever' projects. *Water Sewage & Effluent, 2012*, 18–24.

Tempelhoff, J. W. N. (2009). Civil society and sanitation hydropolitics: A case study of South Africa's Vaal River Barrage. *Physics and Chemistry of the Earth*, Parts A/B/C, *34*(3), 164–175. doi:10.1016/j.pce.2008.06.006

Tewari, D. (2009). A detailed analysis of evolution of water rights in South Africa: An account of three and a half centuries from 1652 AD to present. *Water SA, 35*(5), 693–710. doi:10.4314/wsa.v35i5.49196

UN (United Nations). (2016). Status of the United Nations convention on the non-navigational uses of international watercourses. Retrieved from https://treaties.un.org/Pages/ViewDetails.aspx?src=TREATY&mtdsg_no=XXVII-12&chapter=27&clang=_en (Reference in the article to the 1997 UN Watercourses Convention is to the United Nations Convention on the Non-Navigational Uses of International Watercourses, annexed to UNGA Res. 229 of 21 May 1997, Official Records of the UNGA, 51st session, UN Doc. A/Res/51/229; also reprinted in 36 ILM 700 (1997).

Van Koppen, B., & Schreiner, B. (2014). Moving beyond integrated water resource management: Developmental water management in South Africa. *International Journal of Water Resources Development, 30*(3), 543–558. doi:10.1080/07900627.2014.912111

Van Wyk, J. (2014). Fracking in the Karoo: Approvals required? *Stellenbosch Law Review, 25*(1), 34–54.

Varady, R. G., Van Weert, F., Megdal, S. B., Gerlak, A., Iskandar, C. A., & House-Peters, L. (2012). *Groundwater* policy and governance. Paris: FAO/Global Environment Facility.

Wells, J. D., Van Meurs, L. H., Rabie, M. A., Joubert, G. F., Moir, F., & Russell, J. (2009). Terrestrial minerals. In H. A. Strydom & N. D. King (Eds.), *Fuggle &* Rabie's environmental management in South Africa (2nd ed., pp. 513–578). Cape Town: Juta.

WWF-SA (World Wide Fund for Nature, South Africa). (2015). *Framework to assess the economic reality of shale gas in* South Africa. Claremont: Author.

Groundwater law, abstraction, and responding to climate change: assessing recent law reforms in British Columbia and England

Birsha Ohdedar

ABSTRACT
In 2014, British Columbia enacted the Water Sustainability Act, a comprehensive overhaul of its groundwater and surface water regimes. Meanwhile, in England more piecemeal changes have been made to groundwater laws and policies. Through developing a framework from groundwater governance and climate change adaptation literature this article analyzes the effectiveness of these reforms, which have been carried out through different methods and from different starting points. The article goes on to consider how new processes and technologies, such as hydraulic fracturing (fracking), bring fresh challenges in aligning progress in groundwater law reforms with the wider policy framework.

Introduction

Climate change poses a significant threat to groundwater management around the world. The Intergovernmental Panel on Climate Change (IPCC, 2014) has emphasized that climate change will lead to more extreme events, such as droughts and floods, at more frequent intervals. Groundwater recharge will become increasingly variable and uncertain, which can lead to greater scarcity of water in specific locales, as demand for water also increases. Climate change will require a radical rethink of the assumptions upon which regulations have been made (Craig, 2010). While groundwater issues are immediately critical in countries such as India, as well as arid climates such as Australia and California, traditionally 'wetter' countries such as Canada and England are also feeling the impacts of climate change on groundwater resources.

This article analyzes the effectiveness of law and policy reforms in groundwater abstraction and management in British Columbia (Canada) and England in light of climate change. In particular, it examines British Columbia's new Water Sustainability Act (WSA) in depth, with reforms in England providing a comparison.

Groundwater and climate challenges in British Columbia and England

The relationship between groundwater and climate change in both British Columbia and England is complex and uncertain. There is an overall paucity of research on the relationship, both globally and in the two jurisdictions relevant to this article (Allen, 2009; Jackson, Bloomfield, & Mackay, 2015; Pike et al., 2012). Furthermore, the interactions between groundwater, climate variability, and ecological changes are complex. They can depend on aquifer geography, geology and geometry. The physical aquifer conditions therefore matter greatly and can limit generalized accounts. Nevertheless, the limited research to date in British Columbia and England has identified a number of challenges. These challenges are of course geographically and regionally unique.

Temperature changes and seasonal climate variability will affect groundwater recharge. In British Columbia, interior regions of the province will be particularly sensitive to climate variability, affecting environmental flow and groundwater recharge (Allen, 2009). Similarly, in England, though there is limited research showing systematic changes, there is some evidence of 'multi-annual to decadal coherence' of groundwater levels and large-scale climate indices (Jackson et al., 2015). Research in England has often focussed on particular sensitive regions. For example, Holman and Trawick (2011) have observed the impacts of climate change in East Anglia, noting risks of shorter summer recharge.

A higher occurrence of extreme events due to climate change, such as heavy rainfall, can have an effect on groundwater recharge. Because groundwater is unable to quickly absorb heavy rainfall, there can be greater runoff and flooding. Although these impacts are difficult to quantify accurately hydrologically, they pose a challenge in both jurisdictions (Allen, 2009; Jackson et al., 2015). Extended periods of drought are also a climate risk in specific locales in England and British Columbia. Also, a greater demand for groundwater due to factors related to a changing climate and a higher demand for water, from longer growing season and hotter, drier summers, will increase demand and use of groundwater where surface water is not enough (Allen, 2009).

There is an ongoing challenge for law and policy frameworks in how they understand and incorporate climate and water security risks (Forsyth, 2012; Zeitoun et al., 2016). The many uncertainties and complexities of how socio-ecological systems work are often difficult to translate into simplified policy frameworks (Zeitoun et al., 2016). These points are beyond the scope of this article, but constitute an area of further research and thought. Nevertheless, the trends observed and the research to date fit into broader trends worldwide for an assessment of groundwater governance in light of a changing climate (FAO, 2016; Mechlem, 2016).

Why British Columbia and England?

There are a number of reasons why this article considers groundwater reforms in these two particular jurisdictions. First, both jurisdictions have been active in making reforms in how they regulate groundwater. These recent reforms are thus the focus of the discussion in this article (it is not a thorough primer on groundwater law in each jurisdiction). Second, both jurisdictions are commonwealth legal systems representing a shared method of decision making and water governance. England, of course, represents the source of 'colonial water law' (Simms, Harris, Joe, & Bakker, 2016, p. 6) and the modern riparian doctrine, which many jurisdictions have used as a starting point for regulating groundwater. Nevertheless, while British Columbia has inherited a commonwealth legal system and a colonial water governance system, it has also adopted elements from North American water law, such as 'rule of capture' or 'first in time, first in right' in allocation. The first in time, first in right system gives priority in allocation during times of scarcity to those who hold the most senior licence.

Third, though there is a shared history between British Columbia and England in water governance and legal systems, the two jurisdictions are at very different stages of groundwater regulation. In British Columbia, the government is only just beginning to license groundwater use. The WSA, which received royal assent in 2014 and came into force in 2016, represents a large overhaul and consolidation of the water management and allocation regime in the province. On the other hand, in England, groundwater over-abstraction has become a recognized environmental and social problem, particularly in areas such as the south-east of England, where there is higher demand and relatively less rainfall (Environment Agency, 2013). Thirty-five per cent of groundwater bodies are at risk of not achieving the EU Water Framework Directive's 'good' groundwater quantitative status, which measures the balance between abstraction, recharge and available groundwater.[1]

The government has proposed groundwater abstraction reforms to address the impacts of climate change. For British Columbia, the WSA brings in broad changes to groundwater regulation for the twenty-first century. This provides an opportunity for British Columbia to detach itself from previous eras where there has been a paucity of groundwater regulation. On the other hand, England's regulatory reforms have been done in a more piecemeal manner, providing a contrasting way that groundwater law reform has occurred. Thus the two jurisdictions provide an illustration of the benefits and challenges of reforming groundwater management, from different starting points.

Fourth, both jurisdictions have promoted new technologies and sought sources of energy that have profound affects on groundwater. This article discusses technological advances in relation to hydraulic fracturing ('fracking') for shale gas. British Columbia provides a jurisdiction with a well-developed fracking industry; however, as will be explored, without well-developed or science-based regulatory oversight. This provides potential lessons for England in its governance of fracking in relation to water. Both jurisdictions also highlight the issues of (in)coherence between energy and water policies.

Finally, while this is an examination of two commonwealth jurisdictions and their responses to, and in light of, climate change, it is important to point out that the two countries are different with respect to scale, geography and geology. Nevertheless, in both jurisdictions groundwater has an important role for social, economic and environmental reasons. In England, groundwater plays a particularly important role in the south-east of the country (Environment Agency, 2013). In British Columbia, despite the relative abundance of surface water, groundwater is an important source of drinking water and in some areas is the only viable source of water supply (Wei & Allen, 2004). This article aims to contribute to a more global discussion on groundwater law and climate change through an examination of the recent groundwater law and policy reforms in both contexts, highlighting some insights offered by comparative analysis.

The next section examines how groundwater law and policy can, in general, respond to climate change. The discussion below will then be used as a framework to examine British Columbia's WSA and groundwater abstraction reforms in England.

Responding to climate change and the role of groundwater law

Groundwater law in context

In both jurisdictions, groundwater has taken a secondary role to surface water in water law and policy. This is consistent with trends around the world. As Mechlem (2012, p. 5) notes, 'Historically water legislation has focused on surface water resources, among other reasons because the state of groundwater is unseen, the resource is ubiquitous and aquifer systems respond over time creating less immediate regulatory pressures. Groundwater legislation has lagged behind. In many countries it remains fragmented, incoherent or simply ignored.'

Groundwater law and governance in light of climate change

The effects of climate change on water management and specifically on groundwater management has become an increasingly important issue in water law scholarship (Dellapenna, 2010; Keessen & van Rijswick, 2012; Mechlem, 2016). Water law scholarship has often used the climate change adaptation literature as a framework to argue that particular procedural and substantive concepts must be built into water law (Keessen & van Rijswick, 2012). The broader issues with groundwater globally have also seen a rise in groundwater governance literature, focusing on building the principles and actions necessary to manage groundwater in a sustainable and equitable way. Law plays a central role in groundwater governance through embedding some of its principles in legislative and regulatory regimes (FAO, 2016). For the purposes of this article, we may identify six salient aspects that arise in water law, climate change adaptation and the groundwater governance literature and which may be used to assess recent groundwater law reforms in British Columbia and England.

First, flexibility in water law enables a system to cope with sudden changes in circumstances. Flexibility in water allocation is also important to ensure that water can be reallocated to more valuable uses, from both societal and ecological perspectives (Miller, Rhodes, & Macdonnell, 1997). On the other hand, the rule of law demands certainty, and water users also have historically demanded certainty in their rights. Flexibility also means being able to adapt and change course on the basis of new information (Keessen & van Rijswick, 2012). This also means building in mechanisms to allow regulators to adjust groundwater use rights depending on availability of water, where possible without compensation, and termination on the basis of environmental damage (FAO, 2016). The historical inability of groundwater law to be flexible, through groundwater's being tied to land property rights, is a major reason for multiple water crises today.

Second, participation in information sharing and decision making is a widely recognized element in effective water governance. To enable regulators to make informed decisions, through regulations that are flexible as described above, comprehensive information is critical. Along with data gathering by the government, local knowledge and public participation are key elements of gaining such comprehensive information. Participation is a broad area in the governance of natural resources, and generally, effective public participation in groundwater governance helps ensure legitimacy of decision making and better planning for groundwater goals (Mollenkamp & Kasten, 2009). Nevertheless, for participation to be effective, it requires the continued involvement of the government, in terms of facilitation and financial and institutional support (FAO, 2016).

Third, an effective legal framework is required to deliver groundwater objectives (Keessen & van Rijswick, 2012). Various factors can contribute to an effective legal framework. The key point however is that legal frameworks operate in a way to facilitate climate change mitigation and adaptation, rather than hamper it. This means effective enforcement of laws through responsive and well-resourced regulators. As the Global Framework for Action to Achieve the Vision on Groundwater Governance states, no matter how strong laws and regulations are on paper, 'it is their acceptance, implementation, administration, and enforcement that eventually make the difference' (FAO, 2016, p. 52). Accordingly, 'the capacity of government officials, of local users and of potential polluters to internalize the prescriptions and directions of the law is critical to the ultimate effectiveness of governance arrangements, and must be carefully nurtured' (p. 52).

Fourth, it is vitally important for groundwater law to recognize the connection between surface water and groundwater systems, as well as between groundwater and environmental flow. Historically, surface water and groundwater have been considered separately. Environmental flows describe the timing and amount of water to be retained in lakes, rivers, streams and estuaries to sustain the seasonal patterns of high and low water levels needed for natural functions, processes and resilience to persist (Kendy, Apse, & Blann, 2012). As surface water and groundwater are interconnected, environmental flows affect groundwater.

The relationships between abstractions, groundwater level and river flow are often complex. There is a long lag time in groundwater systems; hence often by the time a policy response is made, the impacts on groundwater systems are already felt (Dyson, Bergkamp, & Scanlon, 2003). This also highlights the need for flexibility in regulation and an effective legal regime. Modifications to environmental flows through human activity affect the functioning of aquifers. Thus, allocation and use of water by competing uses must consider environmental flows to ensure the sustainability of aquifers. Climate variability requires careful planning and decision making around environmental flows, as well as conceptually linking groundwater and surface water to ensure that laws and regulations are in line with ecological baselines.

Fifth, protection of groundwater quality is essential. The impacts of climate change, such as temperature changes in water, can reduce water quality. Furthermore, floods, droughts and other impacts also increase the risk of water pollution. Protecting water quality requires monitoring, as well as other legislative tools, such as setting quality targets in relation to various water uses, classifications of water bodies, reducing and regulating abstraction, prohibitions and limitations on emissions of certain substances, permitting of wastewater discharges, and land-use rules to control 'non-point' sources of pollution (Mechlem, 2016). Along with groundwater law, broader criminal, civil and administrative law can play an important role in regulating the protection of groundwater quality.

Finally, it is essential that linkages are made between groundwater and other areas of law and policy. Groundwater is physically connected to a wide range of human activities, and the historical division of groundwater from society has been a major cause of current water crises. There is an urgent identified need for law and policy in different sectors to align under common principles of climate change adaptation and groundwater governance (FAO, 2016). This article will consider this final point in relation to the link between new technologies, energy and water. It will explore how British Columbia and England's water laws and policy reforms align with corresponding fracking policies. Energy and water policies have historically been looked at in isolation, and this lack of attention is problematic, as any constraint on one is intrinsically linked to the other (King, Stillwell, Twomey, & Webber, 2013).

With the above discussions and framework in mind, we now examine the reforms through the WSA in British Columbia, as well as recent law reforms in groundwater abstraction and management in England.

Recent reforms in British Columbia

Background

In 2014, after a four-year process that included thorough consultation, discussion papers, policy proposals and a proposed legislative framework, the provincial legislature in British Columbia passed the WSA. The WSA is a major overhaul of the water regime in the province. The drivers of reform were population growth, increased water demand, changes in land use, and climate change (British Columbia Ministry of Environment, 2010).

For the first time in its history, groundwater users in British Columbia must now obtain a licence and must pay fees, with the exception of individual household wells, which will not be licensed or charged (WSA, §6). Under the WSA, groundwater licences will be issued to all existing and new users of non-domestic water. Licences are attached to land, and a limited number of activities qualify for licences (§9). Licence holders may use water for conservation, industrial, domestic, irrigation, land improvement, mineral-water extraction, mining, oil and gas, power, storage and waterwork construction purposes. Licensed users must make 'beneficial use' of the water, and if they fail to do so for three consecutive years their licence may be cancelled (§30). Those using groundwater for domestic purposes do not need to apply for a licence, but there is flexibility in the WSA to allow licensing for domestic users through area-based regulations, which could be important in zones of significant water shortages (§136).

Groundwater rights, licences, fees and allocation

As stated earlier, WSA maintains the allocation priority based on first in time, first in right, which has operated to date. This means that if there is a conflict in water use, especially in a situation of water scarcity, the oldest rights are protected first over junior rights. Existing groundwater users have a three-year transition period to seek a licence based on their historic date of first use and their ongoing use of groundwater for non-domestic purposes (domestic use is exempt from requiring a licence).[2] Accordingly, existing well users have until 2019 to submit evidence of their historic use of groundwater to gain senior licence priority. Applications after this date will be treated as new licences.

From a climate change perspective the upholding of first in time, first in right is problematic because it is a priority allocation based solely on being the first to extract. Such a method of allocation is based on principles of politics and power, on who is 'first', rather than being in line with any sustainability or equity consideration (Singh, 1991). The Government of British Columbia (2015a) estimates that there are about 20,000 existing groundwater wells that supply groundwater for non-domestic uses. Regularizing these wells, which could mean up to 20,000 new licences, all with senior rights based on historical first use, and assessing the cumulative impacts of these existing groundwater uses (as well as future uses) on both groundwater and surface water systems will be a major challenge.

In issuing groundwater licences for existing and new uses, the regulator must consider impacts on environmental flows. Interestingly, this was initially not included in the regulations (Government of British Columbia, 2015a). There is very little else in the WSA and associated regulation that guides the terms and conditions that may be applied to meet such a challenge. The risk remains that such licences will be entrenched with seniority based on historical first use, rather than adequately considering impacts on future climatic stresses and demands.

Two aspects of the WSA counteract some of the worst-case scenarios of first in time, first in right. First, where there is a risk of an area's falling below critical environmental flow thresholds, a temporary protection order can be issued to stop abstraction (§§86–-88). Second, there is a 'no compensation' provision, according to which water rights holders are not entitled to any compensation if there is a change, restriction or prohibition on the exercise of the rights, or any imposition of new terms and condition on an approval (§121). This is important because it reduces the substantial monetary concern that may affect a government decision in altering a water right because of unsustainable abstraction. Therefore, while the first in time, first in right model is maintained, the government has room for manoeuvre depending on varying climate-related scenarios.

The fees and charges associated with groundwater licences remain a concern. The charges and fees are relatively low for industrial users, highlighting again the importance of appropriate policy decisions to match the potential of the WSA to deliver good groundwater governance. This has raised a great deal of public concern (Woo, 2015). Undercharging industrial groundwater use would significantly undermine the effectiveness of the licensing regime in British Columbia. In recent years there have been significant concerns about companies such as Nestlé extracting millions of litres of water for free, even during drought conditions (Woo, 2015). While a discussion of the pricing of groundwater is beyond the scope of this article, the important point is that adequate pricing is important to manage groundwater abstraction sustainably and ensure that the correct price signals are given to industry.[3] This includes, for example, regionally differentiated abstraction charges (which are used in England), taking into account regional scarcity. The effectiveness of the licensing regime in British Columbia would be significantly undermined by inappropriate charges; however, as the WSA is so recent, the verdict on whether it can deliver adaptive groundwater management will only be realized in time.

Participatory governance and consideration of environmental flows

Another important reform in the WSA has been the inclusion of Water Sustainability Plans (§§64–85). This reform strengthens the existing framework for water planning and allows the government to delegate responsibility for planning and implementation of groundwater governance. The power given to do this is extensive, and opens up possibilities for inclusive and participative water management. The ability to delegate is wide-ranging under section 126 of WSA, where powers can be transferred to 'another person or entity' to perform the duties instead of the 'comptroller, water manager, engineer or officer'. The effectiveness of this reform hence will depend on a responsive regulator, that delegates to appropriate authorities, with locally sensitive participation.

Effective participation will require more detail on the roles and responsibilities of delegated authorities, and further regulation or Water Sustainability Plans could provide this. It is also important that such participative and collaborative water governance does not effectively see a retreat of the state and is based on delegation to authorities that will have appropriate stewardship responsibilities. On the other hand, the possibilities of collaborative and inclusive water governance mean that First Nations governments, in particular, can play a stronger role in this issue in the province.[4]

The WSA also considers environmental flows in relation to groundwater. The WSA defines environmental flows as 'the volume and timing of water flow required for the proper functioning of the aquatic ecosystem of the stream' (Part 1). This definition has been criticized for being narrower than widely accepted definitions of environmental flow, which consider more than merely the river's hydrology and address water quality and other aspects (Brandes, Carr-Wilson, Curran, & Simms, 2015). Nevertheless, a number of requirements and powers also go towards better protection of environmental flows. These include the ability of a minister or the cabinet to declare a significant water shortage and the comptroller to make a critical environmental flow order (§87), and the competence of the cabinet to establish 'objectives' for the purpose of sustaining water quantity, water quality and aquatic ecosystems (§43).

Such 'objectives' are in fact an important new mechanism to protect water quality, as well as linking groundwater to other activities, and their associated laws and policy. The WSA sets out an enabling provision for 'water objectives' (§43). This is a new mechanism that would allow decision makers to consider the impacts on water when making land and other resource-related decisions. Importantly, these water objectives must link to water quality, water quantity and ecosystem health. Under the WSA, regulations can be made to require public officers making decisions which affect groundwater to consider such water objectives (§43(2)). Furthermore, regulation could also require regional bodies, municipal bodies and local trusts to consider specific water objectives when developing, amending or adopting official growth strategies or community plans (§43(5)). Accordingly, these provisions aim at ensuring that government planning and decision making are consistent with water objectives.

Improving information and knowledge on groundwater

Groundwater data in British Columbia are currently extremely poor and have been found 'insufficient to enable it to ensure the sustainability' of water resources (Office of the Auditor General of British Columbia, 2010, p. 2). The WSA makes efforts to improve this through increasing participation and transparency in data collection. First, there are obligations on licence holders to submit quantitative and qualitative data (§§ 57, 15). Second, water users are required to provide additional detailed monitoring and reporting information in water-scarce areas through Water Sustainability Plans or area-based regulations (§17). Data submission is not mandatory for domestic users, but they will be encouraged to register their wells. If a specific area becomes particularly water-scarce, however, the regulator has the power to enact regulations which require domestic users to also provide such data (§136(1)). Hence, the WSA builds in the ability of the regulatory agency to be flexible in changing conditions. Overall, though information and data will remain a key gap in groundwater governance for British Columbia for a while, the WSA makes significant strides to set up a framework to gather information. In the long term, data will be vital to making governance decisions in light of climate change.

The next section briefly discusses the background of groundwater abstraction and management in England, and then discusses how recent reforms provide a contrast to the major overhaul of groundwater governance in British Columbia.

Recent reforms in England

Background

In England, the main aspects of the system for managing abstraction of water from aquifers were introduced through the Water Resources Act of 1963. Most abstractors were given a licence to extract a fixed volume of groundwater, regardless of availability and with no assessment of environmental impact. Licences were not issued on any scientific basis, as it was assumed that water was a free and plentiful resource. They were issued as the licence applications came in (Sowter & Howsam, 2008). Allocated volumes were based on amounts that had previously been abstracted and on the capacity of abstraction equipment.

Since 2000, reforms of the system have been driven by the EU Water Framework Directive, which has imposed an ecosystem-based approach and has transformed the way water institutions plan and manage groundwater. The Water Act of 2003 brought into focus efficient and sustainable water use, time limitations on new abstraction licences, mechanisms to help licence trading, flexibilities in types of licences, and the deregulation of licensed abstractions of less than 20 m^3 a day. Exemptions were also introduced for certain types of abstraction which were seen to have no significant impact on water quality, and for certain geographical regions. Thus the regime in England represents a hybridized version of the regulated riparian regime (Dellapenna, 2010).

Under the current regime, applications for new licences are assessed with reference to the amount of water available in a particular 'catchment', in accordance with the Catchment Abstraction Management Strategy established in 2001 (a strategy derived from the European Water Framework Directive). All new licences since 2001 are time-limited, mostly to 12 years, but can be renewed as long as they meet tests of environmental sustainability and there is a continued justification of the need for water, as well as efficient use. This system of abstraction is to be reformed through the proposed changes announced in 2016, as discussed below.

Groundwater abstraction reforms since 2011

In 2011, the government released a white paper called *Water for Life* highlighting the issues of over-abstraction, water scarcity, growing demand and climate change, and calling for reform. Since then, the Water Act of 2014 introduced important, yet incomplete, reforms. In January 2016, further abstraction reforms were announced, which signalled an important shift in how groundwater abstraction would be regulated (Defra, 2016a).

Three important changes were enacted through the Water Act 2014: first, it placed a primary duty of 'resilience' and efficient use of water on the Office of Water Services (Ofwat), which is the body responsible for the economic regulation of water (i.e., pricing, investment and management of the privatized water industry in England); second, the act introduced provisions relating to the bulk supply of water, to encourage water trading; and third, it provided for no compensation for water companies whose abstraction licences are varied or revoked on environmental grounds. These changes can be linked to the Environment Agency's programme of 'restoring sustainable abstraction'. Meaningful reform, however, has been postponed until after 2020, with a legislative requirement that a report is tabled in Parliament by May 2019.

Under the 2016 reforms, abstraction licences will be linked to the availability of water and flows, rather than being seasonal or time-limited. Licences will be based on a 'risk-based catchment approach', so that if water availability is limited it can trigger a review if certain thresholds are met. Importantly, all data will be available to the public, to enable abstractors to understand environmental risks and the likelihood of reviews being triggered. Other changes include the continued liberalization of water trading by allowing 'pre-approved' trades, so that permit holders can trade water faster when availability is low. The 2016 reforms are likely to be brought into law in the early 2020s, as part of the wider water abstraction reform.

The government is considering bringing exempted abstractions into the licence control regime (Defra, 2016a). Currently, a number of abstractors are exempt from licence requirements, because their activities are seen as having little impact on water quality. This, together with allowing exempted abstractors to access unlimited water quantities despite water scarcity in given areas, such as areas in the South East of England (Defra, 2016b), has generated a situation where water abstraction is now leading to environmental damage. If brought in, these changes will primarily affect surface water abstractors (who are the majority of water users with exemptions, and will now have to get licences); it is nevertheless a move towards providing for a single system for all water abstractions. It would be in line with the British Columbia regime, which provides one system for all abstractors (except domestic users), with flexibilities, reviews and other mechanisms built in to adapt to varying circumstances and environmental changes.

Effectiveness of changes

The 2016 proposals also signal a strong move towards a more transparent system, which will strengthen the effectiveness of the law to adapt to climate change. The most significant move is the catchment review trigger system, discussed above. The Department for Environment, Food & Rural Affairs (Defra, 2016a) has proposed that the Environment Agency will publish data and information on key indicators in catchments so that abstractors and others are aware of the state of their catchment water assets and of the likelihood that a review could be triggered. The increased transparency will provide certainty for abstractors and flexibility for regulators.

The overall effectiveness of the proposed changes is contingent on a responsive regulator. Unlike in a time-limited approach, reviews are now to be carried out in a more dynamic fashion. Other changes discussed above, such as the power to review and revoke abstraction licences for environmental damage, and the increase in abstractors brought under the licensing scheme, require a responsive and proactive regulator. The law and policy being appropriate to adjust to a changing climate depend on this function. Accordingly, it is important to point out that while the government has been slowly reforming the water abstraction system, it has also reduced funding for the department responsible for regulating water abstraction. According to some analysis, Defra has had the largest proportionate cuts of all government departments in real terms (Howard, 2015). While this article has focussed on legal reform, it is important to point out that reducing the capacity of the regulator is likely to significantly reduce the effectiveness of those laws and policies. In effect, such wider policies have the effect of creating climate vulnerability.

Discussion

In British Columbia, the WSA is robust and progressive legislation for groundwater management. It exhibits most of the characteristics identified earlier in this article for a legal framework that is able to respond to the challenges and uncertainties of a changing climate. A number of issues still remain, however. Foremost among these is how it balances ecological and social considerations with its first in time, first in right priority allocation system, particularly for historical uses that can apply for seniority until 2019. Terms and conditions of such licences, as well as careful consideration of cumulative effects on aquifers, will be important to ensure that the scope for reforming groundwater abstraction that the WSA has provided is not neglected.

In England, on the other hand, robust reforms have essentially been postponed to 2019. The reforms that have happened to date have seen small changes consisting of the introduction of environmental principles into a water abstraction regime built more on property rights than anything else. The removal of compensation for licences deemed to carry out environmentally damaging activity is significant in trying to address this.

In both jurisdictions, the legal framework requires a responsive regulator. In British Columbia, the challenges for a regulator are to use the flexibility and water management tools that the WSA provides – in particular, the ability to have comprehensive water sustainability plans in sensitive areas, such as those in inner British Columbia, which is susceptible to drought (Natural Resource Canada, 2016), and north-eastern British Columbia, where there is more shale gas exploration (Government of British Columbia, 2015b). Similarly, in England, enforcement will become an important element of groundwater management. The proposals to have a dynamic review system also shift the licensing regime towards one based on principles of environmental integrity. Nonetheless, given the long-recognized issue of over-abstraction, reforms in England seem to be moving a lot slower than what is required to keep up with climate change.

Increasing transparency and promoting the development of comprehensive information have become important priorities for both jurisdictions. In British Columbia, this is a 'catch-up' exercise. While the government will publish a five-year report on water in the province, transparent records that have trigger mechanisms to review licences for particular regions could generate a system that may be used in particularly sensitive areas in the future. These mechanisms could be developed through the water sustainability plans. In England's case, the government has maintained information on groundwater for far longer, and it is now developing greater transparency, such as through publicly available information linked to a dynamic review system. Much can be learned from both jurisdictions.

Coherence of law and policy responses: groundwater and regulating fracking

Why explore the role of fracking?

While both jurisdictions have made the reforms discussed above, with an eye on climate change challenges, further issues come from a disjuncture between groundwater regulation and wider economic and energy governance. To explore an example of this, this section will discuss fracking and the regulation of fracking in light of groundwater use. The coherence of groundwater governance and laws and regulations that promote fracking is of vital importance in light of climate change, to ensure that efforts made to reform water law and policies, discussed above, are not undermined through unsustainable practices of the fracking industry. The purpose of this section is to illustrate the importance of such alignment. As mentioned earlier, there is an urgent need for law and policy from different sectors to align under common principles of climate change adaptation and groundwater governance (FAO, 2016). Hence, this section illustrates the challenges that lie in aligning water legislation that is based on principles of good groundwater governance with wider regulation of the energy industry.

Relationship between fracking, groundwater and climate change

The relationship between fracking, groundwater and climate change is multidimensional. Shale gas has gained importance in recent years because it has been promoted as a 'transition fossil fuel' to reduce carbon emissions. As part of their commitments to reduce carbon emissions, governments have been looking towards new forms of energy. Renewable energy, nuclear power, carbon capture and storage, and so-called transition fossil fuels such as shale gas are among the methods countries are using to reduce their emissions. In the US in particular, shale gas exploration through fracking has caused an energy boom and has been seen as a powerful tool for reducing carbon emissions. Both England and Canada have included shale gas as part of their current and future energy mix, but concerns over fracking as a risk to groundwater quality and quantity ('America's falling carbon-dioxide emissions', 2012) have also affected the carbon-mitigation rhetoric and policy choices of governments.

Fracking poses issues for both water quality and quantity. Fracking involves the extraction of oil and gas from 'shale' rock formations by injecting fluids into the earth at high pressure (Department of Energy and Climate Change, 2014). Estimates for how much water a typical drilling operation uses vary between 2 million and 8 millions gallons (Department of Energy and Climate Change, 2014; Flatt & Payne, 2014). Importantly, the vast majority of this water is entirely removed from the hydraulic cycle because it either stays in the formation where it was injected or is introduced into a waste-disposal well (Cooley & Donnelly, 2012; Flatt & Payne, 2014). The high volume of water removed can put considerable stress on water resources at the local level, especially in areas already under scarcity (Wood, Footitt, Nicholls, & Anderson, 2011). Groundwater contamination is also an identified risk in fracking, which has important public health ramifications (Department of Energy and Climate Change, 2014).

British Columbia: different rules for the fracking industry?

In British Columbia, the Oil and Gas Commission (OGC) rather than the Ministry of Environment has an important role in regulating the fracking industry's water use. This is because the industry has widely used short-term water-use approvals under Section 8 of the Water Act of 1996 (henceforth 'Section 8 approval'). The OGC has delegated authority to administer the Water Act in relation to the oil and gas industry. Through this power, the OGC has engaged in a practice of issuing Section 8 approvals recurrently. Thus, the OGC has effectively turned what was intended to be a short-term approval into a medium- and longer-term approval for industry by issuing consecutive short-term approvals. Environmental groups unsuccessfully challenged such recurrent issuing of Section 8 approvals in the British Columbia Supreme Court in 2014 (*Western Canada Wilderness Committee v British Columbia (Oil and Gas Commission)*, 2014 British Columbia SC 1919). The WSA, which succeeds the Water Act, now clarifies that a short-term water use approval can be renewed for up to 24 months. Thus, the fracking industry will continue to use such approvals through the OGC.

The OGC was created in the late 1990s to specifically support the development of the extraction industry. The Ministry of Environment oversees groundwater management in the province more generally. Under the regulations of the Oil and Gas Activities Act of 2008, the OGC must consider the 'government environmental objectives' regarding water, including not allowing the operating area to be located in an identified groundwater recharge area, or on top of an identified aquifer, unless there is no material adverse effect on the quantity, quality or natural flow of water (Environment Protection Management Regulations, Oil and Gas Activities Act, Clause 4). Users are also required to record the amount of water actually withdrawn under Section 8 approvals as well to as disclose chemicals used in the hydraulic fracturing process, and this information is now publicly available (Campbell & Horne, 2011).

While the considerations above do make some linkages between groundwater protection and fracking processes, a wider concern is that of coordination and harmonization between the OGC and the Ministry of Environment. As the licensing regime for groundwater currently stands, a significant portion of groundwater use will remain regulated by the OGC through Section 8 approvals. The OCG has suffered criticism over its effectiveness in enforcing environmental regulations. Collection of the data necessary for informed decisions remains a large gap for the OGC. Section 8 approvals do not set any limit on the amount of water pumped per day, either absolute or relative, and guidelines on the subject are not provided in the manual for the regulator (British Columbia Oil & Gas Commission, 2014).

There is an urgent need for the OGC to have strong guidelines based on basic principles of environmental law, rather than economic efficiency. As a single-window system for the fracking industry, there is a considerable amount of power in a single authority to both promote oil and gas activity and to enforce environmental regulation and oversight. This is often criticized as an inherent institutional problem for the OCG

(Campbell & Horne, 2011). More specifically, there are harmonization issues in having different regulators for groundwater depending on the activity (Ernst & Young, 2015). This has seen, for example, a disjuncture between instances of non-compliance and enforcement – that is, despite numerous instances of non-compliance by the industry, enforcement levels remain relatively low (Hoekstra, 2013).

England: fracking and greying groundwater governance

Fracking for shale gas in England is still in the exploratory phase. Shale gas exploration has been explicitly mentioned as part of the government's plans to decarbonize (Department of Energy and Climate Change, 2015). Despite streamlining several functions in the permitting process for shale gas operation in the UK, groundwater permitting and licensing will still be through the Environment Agency. Therefore, fracking processes will be subject to the same environmental permitting regime for groundwater abstraction as other industries and users.

Recent reforms have raised concern around protection of sensitive groundwater aquifers from fracking. The Infrastructure Act of 2015 brought in a range of reform for large-scale infrastructure projects, including fracking. In the Infrastructure Act, an environmental safeguard prohibits fracking in 'protected groundwater source areas', but the term is not defined. It is expected that a definition will be provided in follow-up regulations. Nonetheless, this was a political U-turn from an earlier proposal to prohibit fracking altogether in key 'groundwater source protection zones' (Friends of the Earth, 2015). Groundwater source protection zones are legally defined as sources of groundwater that are close to drinking water, and thus have public health sensitivity to contamination.

Fracking and the challenges for groundwater governance

Aligning energy policies with groundwater governance needs is still an issue in British Columbia and England. In British Columbia, by enabling an authority other than that responsible for water resources to govern groundwater (the OGC), the legal framework effectively reduces the ability of the Ministry of Environment to make use of the groundwater-related provisions of the WSA. There is an urgent need for governance of groundwater exploitation by the fracking industry to align with the principles of the WSA. In England, political decisions of Parliament have prevented certainty from an environmental point of view as regards fracking activities in sensitive groundwater areas. Overall, in both jurisdictions, law and policy for regulating groundwater use by the fracking industry are not aligned with principles of good groundwater governance and climate change adaptation. This is primarily because governments have made these decisions based on narrow political and economic considerations.

Conclusions

The recent reforms in groundwater law British Columbia and England represent steps towards addressing the challenges of a variable and changing climate. The WSA in British Columbia represents progressive legislation that was long overdue in the province. Groundwater abstraction regulation in England presents a number of innovations, such as catchment based licensing, seasonal licences, and the proposed dynamic review systems, which could also be built into the WSA framework in the future.

In both jurisdictions, certain trends can be observed. Moving beyond historical groundwater rights, granted without regard to social or environmental considerations, is a major challenge in both jurisdictions. Balancing the security of groundwater rights tenure with the challenges of climate variability will continue to be an issue in British Columbia with its 'first in time, first in right' groundwater rights allocation. The various flexibilities and legislative tools that are in the WSA will thus be imperative in future climate and water scenarios.

In both jurisdictions, the importance of a responsive regulator has been highlighted. Proper financial support will be required for the regulator to function effectively. Finally, in both jurisdictions, political decisions around energy and fracking have the potential to undermine the broader groundwater law and policy reforms. There is a need therefore for greater coordination to ensure that fracking does not exacerbate groundwater depletion and degradation. In policy responses to climate change, energy and water must be considered together.

Notes

1. For more information on the EU Water Framework, see European Commission (2016).
2. A significant and long-neglected issue is the conflict between First Nations rights and the first in time, first in right system. This article does not discuss this issue. In British Columbia, the government missed an opportunity to include indigenous peoples' water rights explicitly in the WSA. This omission may create future legal and operational issues (Brandes & Curran, 2016).
3. For more on water pricing and sustainability, see Sjödin, Zaeske, and Joyce (2016) and Rogers, De Silva, and Bhatia (2002).
4. For more information on indigenous water governance in British Columbia and Canada, see Simms (2015) and Bradford, Ovsenek, and Bhardwaj (2016).

Acknowledgements

This article benefited from two workshops in London as part of the partnership project on groundwater and climate change jointly carried out by the National Law University Delhi and SOAS, University of London, under the UK India Education and Research Initiative. I would like to thank all participants at these workshops for their helpful comments and feedback: in particular, Raya Stephan, Philippe Cullet, Andrea Kesseen, Bob Harris and the anonymous referees of the journal. All errors remain mine.

Disclosure statement

No potential conflict of interest was reported by the author.

References

Allen, D. M. (2009, May). Impacts of climate change on groundwater in British Columbia. *Innovation*, pp. 30–32.

America's falling carbon-dioxide emissions: Some fracking good news. (2012, May 5). *Economist.* Retrieved from http://www.economist.com/blogs/schumpeter/2012/05/americas-falling-carbon-dioxide-emissions

Bradford, L., Ovsenek, N., & Bhardwaj, L. A. (2016). Indigenizing water governance in Canada. In S. Renzetti & D. P. Dupont (Eds.), *Water policy and governance in Canada* (pp. 269–300). St. Catharines, ON: Springer.

Brandes, O., Carr-Wilson, S., Curran, D., & Simms, R. (2015). Awash with opportunity ensuring the sustainability of British Columbia's new water law. Retrieved from http://poliswaterproject.org/awashwithopportunity

Brandes, O., & Curran, D. (2016). Changing currents: A case study of the evolution of water law in Western Canada. In S. Renzetti & D. P. Dupont (Eds.), *Water policy and governance in Canada* (pp. 45–69). St. Catharines, ON: Springer.

British Columbia Ministry of Environment. (2010). British Columbia Water Act modernisation technical background report. Retrieved from http://www.livingwatersmart.ca/water-act/docs/wam_tbr.pdf

British Columbia Oil & Gas Commission. (2014). Short-term use water application manual. Retrieved from http://www.bcogc.ca/node/6040/download

Campbell, K., & Horne, M. (2011). Shale gas in British Columbia risks to British Columbia's water resources. *Pembina*. Retrieved from https://www.pembina.org/reports/shale-and-water.pdf

Cooley, H., & Donnelly, K. (2012, June). Hydraulic fracturing and water resources: Separating the frack from fiction. *Pacific Institute*. Retrieved from http://www2.pacinst.org/wp-content/uploads/2013/02/full_report35.pdf

Craig, R. K. (2010). 'Stationarity is Dead' - long live transformation: Five principles for climate change adaptation law. *Harvard Environmental Law Review, 34*, 9–73.

Defra. (2016a). UK government response to consultation on reforming the water abstraction management system. Retrieved from https://www.gov.uk/government/uploads/system/uploads/attachment_data/file/492411/abstraction-reform-govt-response.pdf

Defra. (2016b). Changes to water abstraction licencing exemptions in England and Wales: New authorisation. Retrieved from https://consult.defra.gov.uk/water/water-abstraction-licensing-exemptions

Dellapenna, J. W. (2010). Global climate disruption and water law reform. *Widener Law Review, 15*, 409–541.

Department of Energy and Climate Change. (2014). Fracking UK shale: Water. Retrieved from https://www.gov.uk/government/uploads/system/uploads/attachment_data/file/277211/Water.pdf

Department of Energy and Climate Change. (2015, May 13). Faster decision making on shale gas for economic growth and energy security. Retrieved from https://www.gov.uk/government/news/faster-decision-making-on-shale-gas-for-economic-growth-and-energy-security

Dyson, M., Bergkamp, G., & Scanlon, J. (Eds.). (2003). *Flow: The essentials of environmental flow*. Gland: IUCN. Retrieved from https://cmsdata.iucn.org/downloads/flow___the_essentials_of_environmental_flow___dyson_et_al.pdf

Environment Agency. (2013). Groundwater protection: Principles and practice (GP3). Retrieved from https://www.gov.uk/government/uploads/system/uploads/attachment_data/file/297347/LIT_7660_9a3742.pdf

Ernst & Young. (2015). Review of British Columbia's hydraulic fracturing regulatory framework. Retrieved from https://www.bcogc.ca/node/12471/download

European Commission. (2016). The EU Water Framework Directive: integrated river basin management for Europe. Retrieved from http://ec.europa.eu/environment/water/water-framework/index_en.html

FAO. (2016). Global framework for action to achieve the vision on groundwater governance. Retrieved from http://www.fao.org/3/a-i5705e.pdf

Flatt, V., & Payne, H. (2014). Curtailment first: Why climate change and the energy industry suggest a new allocation paradigm is needed for water utilised in hydraulic fracturing. *University of Richmond Law Review, 48*, 829–856.

Forsyth, T. (2012). Politicizing environmental science does not mean denying climate science nor endorsing it without question. *Global Environmental Politics, 12*(2), 18–23. doi:10.1162/GLEP_a_00106

Friends of the Earth. (2015, February 5). Government backtracks on measures to protect drinking water from fracking. Retrieved from https://www.foe.co.uk/resource/press_releases/government-backtracks-measures-protect-drinking-water-from-fracking_05022015

Government of British Columbia. (2015a). Licensing groundwater use under British Columbia's Water Sustainability Act. Retrieved from https://engage.gov.bc.ca/watersustainabilityact/files/2016/02/LicensingGroundwaterUse-Web-Copy.pdf

Government of British Columbia. (2015b). Northeast water strategy: Ensuring the responsible use and management of Northeast British Columbia's water resources. Retrieved from http://www2.gov.bc.ca/assets/gov/environment/air-land-water/water/northeast-water-strategy/2015-northeast-water-strategy.pdf

Hoekstra, G. (2013, February 18). B.C. Oil and gas commission lacks 'transparency' on fracking violations. *Vancouver Sun*. Retrieved from http://www.vancouversun.com/technology/Commission+lacks+transparency+fracking+violations/7982077/story.html?__lsa=85a5-937e

Holman, I. P., & Trawick, P. (2011). Developing adaptive capacity within groundwater abstraction management systems. *Journal of Environmental Management, 92*(6), 1542–1549. doi:10.1016/j.jenvman.2011.01.008

Howard, A. (2015, November 11). Defra hit by largest budget cuts of any UK government department, analysis shows. *Guardian*. Retrieved from https://www.theguardian.com/environment/2015/nov/11/defra-hit-by-largest-budget-cuts-of-any-uk-government-department-analysis-shows

IPCC. (2014). Climate change 2014: Synthesis Report. Contribution of working groups I, II and III to the fifth assessment report of the intergovernmental panel on climate change, pp. 1–32. Retrieved from https://www.ipcc.ch/pdf/assessment-report/ar5/syr/AR5_SYR_FINAL_SPM.pdf

Jackson, C., Bloomfield, J., & Mackay, J. D. (2015). Evidence for changes in historic and future groundwater levels in the UK. *Progress in Physical Geography, 39*(1), 49–67. doi:10.1177/0309133314550668

Keessen, A. M., & van Rijswick, H. (2012). Adaptation to climate change in European water law and policy. *Utrecht Law Review, 8*(3), 38–50. doi:10.18352/ulr.204

Kendy, E., Apse, C., & Blann, K. (2012). A practical guide to environmental flow for policy and planning with nine case studies in the United States. Retrieved from https://www.conservationgateway.org/ConservationPractices/Freshwater/EnvironmentalFlows/MethodsandTools/ELOHA/Documents/PracticalGuide Eflows for Policy-low res.pdf.

King, C. W., Stillwell, A. S., Twomey, K. M., & Webber, M. E. (2013). Coherence between water and energy policies. *Natural Resources Journal, 53*, 117–217.

Mechlem, K. (2012). Groundwater governance: A global framework for country action - thematic paper 6: Legal and institutional frameworks. *GEF: Groundwater Goverance: A Global Framework for Country Action, GEF ID 3726*. Retrieved from http://ssrn.com/abstract=2177882

Mechlem, K. (2016). Groundwater governance: The role of legal frameworks at the local and national level—established practice and emerging trends. *Water, 8*, 347–363. doi:10.3390/w8080347

Miller, K. A., Rhodes, S. L., & Macdonnell, L. J. (1997). Water allocation in a changing climate: Institutions and adaptation. *Climatic Change, 35*(2), 157–177. doi:10.1023/A:1005300529862

Mollenkamp, S., & Kasten, B. (2009). Institutional adaptation to climate change: The current status and future strategies in the Elbe Basin, Germany. In F. Ludwig, P. Kabat, H. Van Schaik, & M. Micheal Van Der Valk (Eds.), *Climate change adaptation in the water sector* (pp. 227–249). London: Earthscan.

Natural Resource Canada. (2016). Drought. Retrieved from http://www.nrcan.gc.ca/forests/climate-change/forest-change/17772

Office of the Auditor General of British Columbia. (2010). An audit of the management of groundwater resources in British Columbia. Retrieved from https://www.bcauditor.com/sites/default/files/publications/2010/report_8/report/OAGBritishColumbia_Groundwater_Final.pdf

Pike, R. G., Bennett, K. E., Redding, T. E., Werner, A. T., Spittlehouse, D. L., Moore., R. D., ... Tschaplinksi, P. J. (2012). Climate change effects on watershed processes in British Columbia. In R. G. Pike, T. E. Redding, R. D. Moore, R. D. Winkler, & K. D. Bladon (Eds.), *Compendium of forest hydrology and geomorphology in British Columbia* (pp. 694–747). Retrieved from https://www.for.gov.bc.ca/hfd/pubs/docs/Lmh/Lmh66/

Rogers, P., De Silva, R., & Bhatia, R. (2002). Water is an economic good: How to use prices to promote equity, efficiency, and sustainability. *Water Policy, 4*(1), 1–17. doi:10.1016/S1366-7017(02)00004-1

Simms, R. (2015, May) Indigenous water governance in British Columbia and Canada: Annotated bibliography. Retrieved from https://watergovernance.ca/wp-content/uploads/2015/06/Indigenous-water-governance-annotated-bibliography-final.pdf

Simms, R., Harris, L., Joe, N., & Bakker, K. (2016). Navigating the tensions in collaborative watershed governance: Water governance and Indigenous communities in British Columbia, Canada. *Geoforum, 73*, 6–16. doi:10.1016/j.geoforum.2016.04.005

Singh, C. (1991). *Water rights and principles of water resources management*. Bombay: N.M. Tripathi.

Sjödin, J., Zaeske, A., & Joyce, J. (2016). *Pricing instruments for sustainable water management* (Working Paper No. 28). Stockholm: SIWI.

Sowter, P., & Howsam, P. (2008). The Water Act 2003 and sustainable abstraction. *Journal of Water Law, 19*, 1–6.

Wei, M., & Allen, D. M. (2004). Groundwater management in British Columbia, Canada: Challenges in a regulatory vacuum. In M. Brentwood & S. F. Robar (Eds.), *Managing common pool groundwater resources* (pp. 7–31). Westport, Conn.: Praeger.

Woo, A. (2015, July 14). British Columbia to review proposed groundwater pricing. *Globe and Mail*. Retrieved from *http*://www.theglobeandmail.com/news/british-columbia/bc-to-review-proposed-groundwater-pricing/article25512023/

Wood, R., Footitt, A., Nicholls, F., & Anderson, K. (2011, July). Shale gas: A provisional assessment of climate change and environmental impacts. *Tyndall Centre for Climate Change Research*. Retrieved from http://www.tyndall.ac.uk/sites/default/files/tyndall-coop_shale_gas_report_final.pdf

Zeitoun, M., Lankford, B., Krueger, T., Forsyth, T., Carter, R., Hoekstra, A. Y., ... Matthews, N. (2016). Reductionist and integrative research approaches to complex water security policy challenges. *Global Environmental Change, 39*, 143–154. doi:10.1016/j.gloenvcha.2016.04.010

EU legal protection for ecologically significant groundwater in the context of climate change vulnerability

Owen McIntyre

ABSTRACT

EU habitats law can provide robust protection for groundwater supporting legally protected habitats, such as wetlands. Court of Justice of the EU jurisprudence requires precautionary assessment of groundwater's role in maintaining the 'integrity' of protected ecosystems. Precaution applies in cases of scientific uncertainty, such as that pertaining to groundwater ecology, exacerbated by the uncertain effects of higher climate variability. While EU habitats law may not address anthropocentric concerns, an expansive approach to the protection of groundwater, guided by precautionary assumptions concerning its ecological role, can safeguard essential water-related ecosystem services and thereby address human needs threatened by climate change.

Introduction

While EU law aimed at the protection of groundwater may be regarded as relatively weak, the well-established corpus of EU nature conservation law, which is applied in a robust manner by EU and national courts, offers alternative legal protection. In particular, the concept of ecological 'integrity' employed in the 1992 EU Habitats Directive, and thus in the nature conservation law of each EU member state, can operate to ensure the effective protection of groundwater resources which perform an ecological function for legally protected ecosystems. The concept of 'integrity' occupies an absolutely central place in the field of EU nature conservation law as the key substantive standard of legal protection afforded to sites designated under both the 1979 Wild Birds Directive (replaced by Directive 2009/147/EC) and the 1992 Habitats Directive. It provides an example of a vague and flexible environmental standard requiring further administrative or judicial elaboration for its consistent and predictable application. Recent pronouncements on the concept by the Court of Justice of the European Union (CJEU), most notably in the case of *Sweetman v. An Bord Pleanála* (CJEU, 2013), suggest that the court will take a very robust view of the standard of ecological protection stipulated therein, a view which could extend significant protection to groundwater resources with an ecological function for the protected ecosystem in question.

In its reasoning the court has employed several modes of creative legislative interpretation, each of which suggests that ecologically significant groundwater will enjoy a very high level of legal protection under EU nature conservation law, especially in the light of persistent scientific uncertainty regarding the role of groundwater in the structure and function of protected habitats. Uncertainty regarding the status and hydrological character of groundwater bodies has a number of technical and human causes (UNESCO-IHP, 2015). Such uncertainty is greatly exacerbated by the risks and further scientific uncertainty associated with climate change. While it is perfectly well understood that many groundwater resources may be vulnerable to climate change, for example due to changing patterns of aquifer recharge, there is great uncertainty regarding the potential impacts of climate change on the role of groundwater in the maintenance and function of habitats (UNESCO-IHP, 2012). Of course, in some cases groundwater resources will also have a potential role in responding to the adverse impacts of climate variability by providing resilient 'reserves' where aggravated periods of surface water scarcity threaten habitat integrity (OECD, 2015). Indeed, UNESCO-IHP (2015, p. 3) recently noted that the role of groundwater in responding to climate change impacts might in turn give rise to new risks and further uncertainty: 'Direct impacts of climate change on natural processes (groundwater discharge, recharge storage and quality) may be exacerbated by the human response to these impacts, such as increased groundwater abstraction due to extended and more frequent droughts.'

While EU nature conservation law cannot be employed, at least directly, to safeguard groundwater resources required for purely human needs (ECJ, 1991), the protection of groundwater for the purposes of maintaining protected habitats such as wetlands might also result in improved provision of related anthropocentric ecosystem services, such as water storage, water supply, flood control or recreational uses (McIntyre, 2014; Millennium Ecosystem Assessment, 2005).

EU groundwater law

For many years, a range of EU legislative instruments have sought to address the protection of groundwater, though these have tended to be primarily concerned with the management of groundwater pollution rather than with quantitative groundwater resources management. For example, the 1976 Directive on Discharges of Certain Dangerous Substances Discharged into the Aquatic Environment (76/464/EEC) required member states to take appropriate steps to reduce and eliminate pollution of groundwater by particular substances listed in the directive, and further required the laying down of emission standards and limit values to that end. In addition, a number of directives have sought to address pollution from specific priority substances, such as mercury and cadmium. Further EU controls on the management of hazardous wastes and the handling of hazardous substances may also be indirectly relevant to the protection of groundwater from pollution. The 2000 Water Framework Directive now requires member states to take 'the measures necessary to prevent or limit the input of pollutants into groundwater and to prevent the deterioration of the status of all bodies of groundwater', while also prohibiting 'direct discharges of pollutants into groundwater'. The 2006 Directive on the Protection of Groundwater against Pollution and

Deterioration further details the obligation of member states to prevent direct inputs of hazardous substances and to limit direct inputs of non-hazardous polluting substances, while also taking account of pollutants from diffuse sources.

Regarding the quantitative management of groundwater resources, the Water Framework Directive provides that 'Member States shall protect, enhance and restore all bodies of groundwater, [and] ensure a balance between abstraction and recharge of groundwater, with the aim of achieving good groundwater status'. The directive also unequivocally lists among its objectives that of establishing a framework which 'prevents further deterioration and protects and enhances the status of aquatic ecosystems and, with regard to their water needs, terrestrial ecosystems and wetlands directly depending upon the aquatic ecosystems', as well as that of 'mitigating the effects of floods and droughts'. Article 6(1) of the directive expressly acknowledges the possible links between groundwater and protected habitats by requiring that

> Member States shall ensure the establishment of a register or registers of all areas lying within each river basin district which have been designated as requiring special protection under specific Community legislation for the protection of their surface water and groundwater or for the conservation of habitats and species directly depending on water.

More generally, it recognizes that 'the quantitative status of a body of groundwater may have an impact on the ecological quality of surface waters and terrestrial ecosystems associated with that groundwater body'. Further, regarding the monitoring of groundwater status, Article 8 requires that, for protected areas, the monitoring programmes to be established by member states include coverage of 'those specifications contained in Community legislation under which the individual protected areas have been established'.

Annex V of the Water Framework Directive also elaborates upon 'groundwater quantitative status' so that 'the level of groundwater is not subject to anthropogenic alterations such as would result in ... any significant damage to terrestrial ecosystems which depend directly on the groundwater body'. Annex VII sets out the required content of a river basin management plan, the key instrument for the coordination of administrative arrangements in river basins and for implementing the substantive objectives set out in the directive, and stipulates the 'mapping of ecoregions ... within the river basin, estimation of pressures on the quantitative status of water including abstractions, a list of environmental objectives established ... for surface waters, groundwaters and protected areas ... [and] details of the supplementary measures identified as necessary in order to meet the environmental objectives established'.

Thus, the Water Framework Directive, the primary legislative instrument for the management and protection of water resources across the entire EU, is clearly intended to protect groundwater and maintain its function for protected habitats and ecosystems. This function of groundwater is increasingly clearly understood (FAO 2015). In addition, as it is also concerned with mitigation of the impacts of floods and droughts, it is reasonable to expect that it should aim to do so in light of climate change impacts. However, it does not appear that it is operating in practice to meet this objective. A 2007 EU Commission study has found that over 40% of river basin management plans don't address the issue of water scarcity and drought at all, while only 12% identify water scarcity and drought pressures by sector, and only 5% include coordinated measures to address this concern, which is increasingly widespread throughout

Europe, largely due to impacts associated with climate change (EU Commission, 2007). A 2012 commission review of progress on this issue concluded that trends in water scarcity and drought management in Europe have not been reversed, primarily for lack of a dedicated legal instrument which, taking account of climate change risks, might define and implement ecological flow requirements and water efficiency targets, promote economic incentives for water efficiency and guide land-use practices to respond to water scarcity (EU Commission, 2012a). While the commission has also adopted a *Blueprint to Safeguard Europe's Water Resources* (EU Commission, 2012b), which places considerable emphasis on land-use practices and their impact on the ecological status of water resources and seeks to promote assessment and management of water scarcity and drought specifically to mitigate the effects of climate change, it is not at all clear how this can be achieved in the absence of a new legislative framework. Notably, the blueprint calls for the integration of water quantity issues more fully into the overall policy framework, presumably including the EU nature conservation regime, a requirement that is likely to become ever more urgent as climate change impacts become more fully understood.

Ecological 'integrity' in EU nature conservation law

EU nature conservation law, a relatively discrete body of rules including the 1979 Wild Birds Directive (as replaced by Directive 2009/147/EC) and the 1992 Habitats Directive, pursues an 'enclave' strategy, requiring the active designation of areas enjoying special nature conservation status, within which special rules of environmental protection are to apply (Scott, 1998). Notwithstanding the inclusion of provisions on the protection of species beyond designated sites, the key means of legal protection of habitats and species in EU law involves preventing national authorities from approving plans or projects which might adversely affect sites of high ecological value designated under either directive, which are collectively known as Natura 2000 sites. To this end, Article 6(3) of the Habitats Directive is the key provision, requiring that an 'appropriate assessment' be carried out for any plan or project which might significantly affect a Natura 2000 site. It is not difficult to imagine a plan or project, such as a proposal for a natural gas 'fracking' project, which might harm groundwater resources playing an important ecological role in relation to a protected site. Article 6(3) provides that

> Any plan or project ..., either individually or in combination with other plans or projects, shall be subject to appropriate assessment of its implications for the site in view of the site's conservation objectives. In the light of the conclusions of the assessment of the implications for the site ... the competent national authorities shall agree to the plan or project only after having ascertained that it will not adversely affect the *integrity* of the site concerned.

Therefore, although Article 6(4) provides exceptions to the rule in Article 6(3), whereby plans or projects which have been found to present a risk to the integrity of the site may be permitted on grounds of 'imperative reasons of overriding public importance', the appropriate assessment required under Article 6(3) represents the single most important legal mechanism for the protection of European habitats and species (Scott, 2012). Unlike environmental impact assessment or strategic environmental assessment, an appropriate assessment of the effects of a plan or project on a Natura 2000 site is determinative of the outcome of the approval process, and this assessment must make a determination on the basis of a single objective substantive standard, i.e., that of maintenance of 'the integrity of the site concerned'. Thus, the scientific and legal nature of the 'integrity' standard is of central concern to the effective implementation of the 1979 (2009) and 1992 directives, which form the principal pillars of EU nature conservation law.

The 'appropriate assessment' process

Official commission guidance on the conduct of an appropriate assessment sets out in considerable detail the precise nature of each step required and their sequence (EU Commission, 2002). It stipulates four distinct stages:

(1) Screening – to determine whether there are likely to be any significant effects on a Natura 2000 site.
(2) Appropriate assessment – to determine whether there will be any adverse effects on the integrity of a Natura 2000 site.
(3) Assessment of alternative solutions – to determine whether there are any alternatives to the proposed project or plan, where it is likely to have adverse effects on the integrity of a Natura 2000 site.
(4) Assessment of compensatory measures – to determine whether there are compensation measures which could maintain or enhance the overall coherence of Natura 2000, where the plan or project is likely to have adverse effects on the integrity of a Natura 2000 site.

Only the first two stages are of interest for this article, as Stages 3 and 4 only become relevant where the plan or project as originally proposed has already been found to be likely to damage the integrity of a Natura 2000 site and the exceptional provisions of Article 6(4) may apply. Under Article 6(4) such a plan or project may be permitted to proceed if it 'must nevertheless be carried out for imperative reasons of overriding public interest, including those of a social or economic nature'. It is worth noting that, where Article 6(4) is invoked, the European Court of Justice (EJC, 2012, para. 122) has found, in relation to surface water resources, that 'irrigation and the supply of drinking water meet, in principle, those conditions and are therefore capable of justifying the implementation of a project for the diversion of water in the absence of alternative solutions'.

Screening

Stage 1 requires a description of the plan or project in question and of other plans or projects that in combination could have significant effects on the Natura 2000 site, as well as identification of these potential effects and assessment of the extent of their impact with regard to key significance indicators, including the duration or permanence of disturbance to habitats, and *relative change in water resource and quality* (EU Commission, 2002, pp. 18–20, emphasis added). Helpful judicial statements exist on the sequential ordering and intensity of the initial two stages of assessment required under Article 6(3). In the *Waddenzee* case, the ECJ (2004, para.43) explained that

> the first sentence of Article 6(3) of the Habitats Directive subordinates the requirement for an appropriate assessment of the implications of a plan or project to the condition that there be a probability or a risk that the latter will have significant effects on the site concerned.

Therefore, a second, more detailed assessment is required where the preliminary assessment identifies a risk of significant effects with regard to the precautionary principle, 'by reference to which the Habitats Directive must be interpreted' (ECJ, 2004, para.44; Jans & Vedder, 2008). Indeed, the court found that 'such a risk exists if it cannot be excluded on the basis of objective information that the plan or project will have significant effects on the site concerned', which in turn 'implies that in case of doubt as to the absence of significant effects such an assessment must be carried out' (ECJ, 2004, para. 44). Thus, 'the case law of the ECJ makes it clear that the trigger for an appropriate assessment is a very light one, and the mere probability or a risk that a plan or project might have a significant effect is sufficient to make an "appropriate assessment" necessary' (Simmons, 2010, p. 7). This approach to judicial interpretation of the requirement for screening under Article 6(3), in combination with the scientific uncertainty commonly associated with the vulnerability of groundwater, with its role in the maintenance of connected ecosystems, and with the likely effects of climate change on water resources availability, makes it highly likely that any potential impacts on groundwater would trigger the need for a 'full' Stage 2 assessment.

Appropriate assessment

If the screening stage concludes that there may be significant effects on a Natura 2000 site, Stage 2 requires that an appropriate assessment be conducted. It is clear from the commission's guidance on the implementation of Article 6(3) that this assessment of the impact of the plan or project on the integrity of the site involves a structured process consisting of four key steps: gathering of all relevant information; prediction of the likely impacts of the plan or project; assessment of whether these impacts will have adverse effects on the integrity of the site with regard to its conservation objectives and status; and assessment of measures proposed to counteract the adverse effects the project is likely to cause (EU Commission, 2002).

The information to be gathered and considered will involve a range of information about the project, including, for example, the results of any environmental impact assessment or strategic environmental assessment, and a range of information about the site's conservation objectives and status, the key attributes of protected habitats or species on the site, and the key structural and functional relationships that create and maintain the site's integrity (EU Commission, 2002). Clearly the latter might require investigation of the role of groundwater in maintaining the site's ecological integrity and of any likely change in the ecological role of groundwater due to the impacts of climate change. The commission's uidance also lists among the information essential for completion of the Article 6(3) appropriate assessment 'the characteristics of existing, proposed or other approved projects or plans which may cause interactive or cumulative impacts with the project being assessed and which may affect the site' (p. 26).

The EU Commission's (2002) guidance details the range of 'impact prediction methods' which might be employed, including quantitative predictive models, geographical information systems, information from previous similar projects, and expert opinion and judgment. More generally, the guidance stresses that 'predicting impacts should be done within a structured and systemic framework and completed as objectively as possible' (p. 27). The 'existing baseline conditions of the site' is expressly included among the information required under the commission's guidance to complete an appropriate assessment, and 'where [such] information is not known or not available, further investigations will be necessary' (pp. 25–26). This explicit requirement for adequate baseline information might operate to promote thorough scientific investigation of groundwater resources in the vicinity of protected sites. Reliable baseline data on groundwater are often non-existent, and so such investigation would be a vital first step in its sustainable management and effective protection, and imperative in the light of anticipated climate change impacts.

Assessment of whether there will be adverse effects on the integrity of the site as defined by its conservation objectives and conservation status must apply the precautionary principle and involves completion of the 'integrity of site checklist' provided under the EU guidance (p. 28). The checklist asks whether the plan or project delays or interrupts progress towards the conservation objectives of the site, whether it disrupts key factors which help maintain the favourable conditions of the site, and whether it interferes with the balance, distribution or density of key species that are indicators of the favourable condition of the site (p. 28). It also asks whether the plan or project affects a range of other indicators, including vital aspects of the structure and functioning of the site and loss or reduction of key ecological features (p. 29). Therefore, the continuing availability of ecologically significant groundwater might in many cases be regarded as such a key factor, vital aspect or ecological feature, especially considering the central role of the precautionary principle and the aggravated risk of water scarcity due to climate variability.

Regarding the assessment of mitigation measures, the guidance insists that these should aim for the top of the mitigation hierarchy, setting out the preferred approaches to mitigation in this order: 1. avoid impacts at source; 2. reduce impacts at source; 3. abate impacts on site; 4. abate impacts at receptor (p. 14). Clearly, this militates against any plan or project that might threaten the availability or quality of ecologically significant groundwater. The assessment of mitigation measures also involves, initially, listing each of the measures to be introduced and explaining how they will avoid or reduce adverse impacts on the site. Then, for each mitigation measure, it is necessary to provide a timescale of when it will be implemented and to provide evidence of how it will be implemented and by whom, of the degree of confidence in its likely success, and of how it will be monitored and rectified in the event of failure (pp. 30–31). The usual degree of uncertainty concerning the status of groundwater resources, compounded by the uncertain consequences of climate change for such resources, suggest that it would normally be very difficult to provide the requisite assurance that the proposed mitigation would avoid or adequately reduce adverse impacts on relevant groundwater.

Judicial deliberation on appropriate assessment

Though the key issue in any Article 6(3) appropriate assessment is that of whether a plan or project 'adversely affects the *integrity* of the site concerned' (Jones, 2012, p. 151, emphasis added), judicial clarification of this concept had not come along until the recent *Sweetman* judgment. The ECJ had, however, provided some judicial clarification as to the boundaries of the integrity standard and the procedural requirements for an appropriate assessment, causing leading commentators to observe that 'the Court has put the bar quite high indeed' (Jans & Vedder, 2008, p. 461). In the *Waddenzee* case, for example, the ECJ (2004, para. 61) stated categorically that 'The competent national authorities, taking account of the appropriate assessment of the implications ... are to authorize such an activity only if they have made certain that it will not adversely affect the integrity of that site. This is the case where no reasonable scientific doubt remains as to the absence of such effects.'

This level of scientific certainty might be very difficult to achieve in the case of any proposal with the potential to affect important groundwater resources. With regard to the finding in an environmental impact study that 'the project in question has a "significantly high" overall impact and a "high negative impact" on the avifauna present in the Castro Verde Special Protection Area (SPA)', the court in *Commission v. Portugal* found that 'the inevitable conclusion is that, when authorizing the planned route of the A motorway, the Portuguese authorities were not entitled to take the view that it would have no adverse effects on the SPA's integrity' (ECJ, 2006, paras. 22–23). In this case, the court reminded the Portuguese authorities that they had other options under the Habitats Directive for authorizing the project: 'in those circumstances, the Portuguese authorities had the choice of either refusing authorisation for the project or of authorising it under Article 6(4) of the Habitats Directive, provided that the conditions laid down therein were satisfied'. Thus, the availability of the exceptional power to authorize under Article 6(4) might suggest a very high standard of protection under Article 6(3).

In *Commission v. Italy*, the ECJ had to determine whether a 2000 environmental impact study and a further 2002 report were adequate in combination to be considered an appropriate assessment within the meaning of Article 6(3). In reaching the 'inescapable conclusion' that the earlier study did 'not constitute an appropriate assessment on which the national authorities could rely for granting authorisation for the disputed works pursuant to Article 6(3) of Directive 92/43' (ECJ, 2007, para. 65), the court emphasized 'the summary and selective nature of the examination of the environmental repercussions' of the proposed works, the fact that the 'study itself mentions a large number of matters which were not taken into account' and thus recommended 'additional morphological and environmental analyses and a new examination of the impact of the works ... on the situation of certain protected species' and, further, that 'the study takes the view that the carrying out of the proposed works ... must comply with a large number of conditions and protection requirements'. Regarding the later report, the court reached a similar conclusion, and complained that it 'does not contain an exhaustive list of the wild birds present in the area' for which the SPA at issue had been designated, that it 'contains numerous findings that are preliminary in nature and it lacks definitive conclusions', and further, that it stresses 'the importance of assessments to be carried out progressively, in particular on the basis of knowledge and details likely to come to light during the process of implementation of the project'. Indeed, the court provides a very clear and concise indication of the deficiencies in an assessment which would render it inadequate for the purposes of Article 6(3) of the Habitats Directive:

> It follows from all the foregoing that both the study of 2000 and the report of 2002 have gaps and lack complete, precise and definitive findings and conclusions *capable of removing all reasonable scientific doubt as to the effects* of the works proposed on the SPA concerned. (para. 69, emphasis added)

This judicial reasoning suggests once again that very comprehensive data on groundwater baseline conditions and on likely groundwater impacts, which might not be easily available, would be required for approval of any proposal likely to affect ecologically relevant groundwater. Increasing awareness of potential climate change effects on groundwater resources (UNESCO-IHP, 2012) would require that such effects also be considered, making it even more difficult to remove 'all reasonable scientific doubt as to the effects' on the protected site.

Judicial interpretation of ecological 'integrity'

Despite the availability of technical guidance on the implementation of Article 6 at both EU and national levels, and judicial guidance such as that outlined immediately above, considerable uncertainty persisted about the precise meaning of the 'integrity' of a protected site (Opdam, Broekmeyer, & Kistenkas, 2009; Söderman, 2009; Therivel, 2009). Questions regarding the conservation implications of the concept of ecological 'integrity' as pursued under Article 6(3) of the Habitats Directive eventually came before the CJEU in the *Sweetman* case, some 20 years after the adoption of the directive. This case concerned a proposed road project in Ireland, the N6 Galway Outer Bypass, which would have led to the permanent loss of 1.47 hectares of limestone pavement, an Annex I priority habitat type, in the Lough Corrib candidate Special Area of Conservation (SAC), which covers 25,000 hectares. The area to be affected was in a distinct sub-area of the site containing 85 hectares of limestone pavement, out of a total 270 hectares of this particular geological feature found throughout the candidate SAC, which was one of six priority habitat types out of a total of 14 Annex I habitats hosted by the site and recognized as ecologically important in terms of the SAC's conservation objectives. The Irish Supreme Court eventually referred the question of 'integrity' under Article 6(3) to the CJEU – the first time that the Irish courts have done so in a case involving the interpretation of EU environmental law. The Supreme Court requested a preliminary ruling on three closely interrelated questions, which centred on the 'the criteria in law to be applied by a competent authority to an assessment of the likelihood of a plan or project ... having "an adverse effect on the integrity of the site"', and on the related 'application of the precautionary principle ... [and] ... its consequences'.

Following the very robust opinion of the advocate general, the court determined that any permanent loss of the habitat type for which the site had been designated must 'adversely affect the integrity of the site' for the purposes of Article 6(3). Advocate General Sharpston had earlier unequivocally concluded that

> measures which involve the permanent destruction of a part of the habitat in relation to whose existence the site was designated are, in my view, destined by definition to be categorized as adverse. The conservation objectives of the site are, by virtue of that destruction, liable to be fundamentally – and irreversibly – compromised. The facts

> underlying the present reference fall into this category. (Opinion of Advocate General Sharpston, 2012, para. 60)

For the purposes of the present article, it is argued that the modes of reasoning employed by the court (and by the advocate general) in arriving at this conclusion may have far-reaching implications for the application of the concept of ecological 'integrity' for the effective conservation of groundwater resources having an ecological role in Natura 2000 sites.

Teleological legislative interpretation

The court characteristically engaged in teleological interpretation of the relevant provisions of the Habitats Directive, a mode of judicial interpretation by which it interprets legislative provisions holistically in the light of the purpose these provisions aim to achieve (Fennelly, 1996; Maduro, 2007). Therefore, the court agreed with the advocate general that Articles 6 (2–4) of the directive and the scope of the expression 'adversely affect the integrity of the site', which 'must be construed as a coherent whole in the light of the conservation objectives pursued by the directive', and, further, that these provisions 'impose upon the Member States a series of specific obligations and procedures designed ... to maintain, or as the case may be restore, at a favourable conservation status natural habitats and, in particular, special areas of conservation' (CJEU, 2013, para. 36). Thus, pursuant to this mode of legislative interpretation, the court has closely linked the notion of site 'integrity' to the directive's objective of maintaining or restoring important natural habitats and species at a 'favourable conservation status', inferring that 'for the integrity of a site as a natural habitat not to be adversely affected ... the site needs to be preserved at a favourable conservation status'. It cites with approval the advocate general's observation that 'this entails ... the lasting preservation of the constitutive characteristics of the site concerned that are connected to the presence of a natural habitat type whose preservation was the objective justifying the designation of that site' (para. 56). The court therefore confirmed that, for the purposes of Article 6(3), 'the conservation objective thus corresponds to maintenance at a favourable conservation status of that site's constitutive characteristics'. The court also found that this requirement to ensure lasting preservation of key ecological characteristics of a designated site 'applies all the more' where the natural habitat affected 'is among the priority natural habitat types', and concluded that 'the precautionary principle should be applied for the purposes of that appraisal' (para. 48). It is not difficult to imagine many situations where ecologically significant groundwater might be counted among such constitutive or key ecological characteristics connected to a priority natural habitat type, such as a wetland or river.

The court, therefore, has adopted a very strict understanding of the requirement to maintain the ecological integrity of a protected site under Article 6(3), at least regarding likely permanent or long-lasting loss or damage of a priority habitat type, the preservation of which was intended by designation of that site. The issue of permanence or long-lasting effect is central, and the advocate general distinguished from the situation arising in the present case a 'plan or project [that] may involve some strictly temporary loss of amenity which is capable of being fully undone' (Opinion of Advocate General Sharpston, 2012, paras. 59–61). She stated plainly that, '*provided* that any disturbance to the site could be made good, there would not ... be an adverse effect on the integrity of the site' (para. 59). Any sustained interference with the availability of groundwater which functions to support ecosystems would be likely to have a permanent or long-lasting adverse effect on such ecosystems. In addition, when one considers that the court has reaffirmed that 'the Habitats Directive has the aim that the Member States take appropriate protective measures to preserve the *ecological characteristics* of sites which host natural habitat types' (CJEU, 2013, para. 38), it appears that ecologically relevant groundwater will receive a high degree of protection, especially in light of the ecological criteria detailed in the 'integrity of site checklist' provided in the commission's guidance on the implementation of Article 6 (EU Commission, 2002, pp. 28–29).

Precautionary principle

Both the court and the advocate general base this strict interpretation of the requirement to maintain a protected site's ecological 'integrity', at least in part, on the precautionary principle. While the precautionary principle is not expressly mentioned anywhere in the Habitats Directive, the court had already enthusiastically established that Article 6(3) of the directive integrates the precautionary principle (ECJ, 2004, paras. 44 and 58), and now appears to regard the principle as indispensable to that provision's effective implementation. This appears to be an example of the court's use of the *effet utile* (useful effect) doctrine in the interpretation of Article 6(3), described by Judge Fennelly (1996, p. 674) as the 'constant companion of the chosen [teleological] method', which provides that 'once the purpose of a provision is clearly identified, its detailed terms will be interpreted so "as to ensure that the provision retains its effectiveness"'. Though he concedes that it might appear somewhat 'shocking' to the common lawyer that 'the Court either reads in necessary provisions ..., or bends or ignores literal meanings ... [or] ... fills in lacunae which it identifies in legislative or even EC Treaty provisions', Fennelly stresses the Court's use of the *effet utile* doctrine in creating a 'Community of law' extending beyond the merely economic objectives of the early EU Treaties (p. 676). The court can also be understood as employing a 'contextual' variant of the teleological approach, whereby it interprets a provision of EU law by considering 'not only its wording, but also the context in which it occurs and the objects of the rules of which it is a part' (p. 664). Its reasoning provides an example of the precautionary principle performing its 'guidance function', as a guiding principle of EU environmental law, whereby 'European law may – and indeed must – be interpreted in the light of the environmental objectives of the TFEU [Treaty on the Functioning of the European Union]' (Jans, 2010, p. 1541; see also McIntyre, 2012).

Commentators have long understood the precautionary principle as 'a tool for decision-making in a situation of scientific uncertainty', which effectively 'changes the role of scientific data' (Freestone, 1994, p. 211; see also McIntyre & Mosedale, 1997). For the principle's application, therefore, there should exist a state of scientific uncertainty – the normal situation for groundwater and its ecological significance, even before one takes account of further uncertainty regarding the impacts of climate change. Both the advocate general and the court expressly linked the application of the precautionary principle with situations 'where uncertainty remains as to the absence of adverse effects on the integrity of the site'. Such an understanding of the applicability of the precautionary principle accords with the key commission guidance on the principle, which advises that

> application of the precautionary principle is part of risk management, when scientific uncertainty precludes a full assessment of the risk and when decision-makers consider that the chosen level of environmental protection or of human, animal and plant health may be in jeopardy. (EU Commission, 2000, p. 13).

Indeed, scientific uncertainty is rather more likely to feature in cases involving groundwater impacts than it was in the present case, where it was not apparent that there was any real uncertainty as to the existence or extent of the ecological risks involved. The court concluded that 'a less stringent authorisation criterion' than that based on the precautionary principle 'could not ensure as effectively the fulfilment of the objective of site protection intended under that provision'. Scientific uncertainty regarding climate change impacts and their direct and indirect effects on groundwater resources can only greatly exacerbate the uncertainty already associated with the status of groundwater bodies and with their role in sustaining ecosystems. Therefore, the court clearly regards the precautionary principle as capable of informing the substantive standard of protection afforded to a protected site, and to its key ecological components, which would often include groundwater playing an ecologically relevant role.

Policy grounds

The court's reasoning in the *Sweetman* case would also appear to employ elements of what the common lawyer would recognize as a policy decision (Atiyah, 1988; Bell, 1983), going beyond the more usual teleological (purposive) approach, whereby the court may resort to 'other criteria of interpretation, in particular the general scheme and the purpose of the regulatory system of which the provisions in question form part' (Fennelly, 1996, p. 665). It can even be regarded as going beyond the judicial creativity of the *effet utile* doctrine, by means of which 'the Court fills in lacunae which it identifies in legislative ... provisions' (p. 674). Most notably, the advocate general argued that any interpretation of site 'integrity' other than the very strict one advanced in her opinion would fail to prevent 'the "death by a thousand cuts" phenomenon, that is to say, cumulative habitat loss as a result of multiple, or at least a number of, lower level projects being allowed to proceed on the same site'. Though she also insisted that this phenomenon has no role 'in determining whether the "adverse effect on the integrity of the site" test under Article 6(3) was met' (Opinion of Advocate General Sharpston, 2012, para. 67), this reassurance is not entirely convincing. For example, she elsewhere criticized the contrary view of the meaning of integrity on the basis that 'it ... fails in any way to deal with the "death by a thousand cuts" argument' (para. 74). Though the court did not refer explicitly to this phenomenon, it implicitly supported the advocate general's concerns. Such an approach can only advance stringent protection of ecologically relevant groundwater resources with regard to their inherent vulnerability and to the risks presented by climate change. For example, on the basis of this reasoning any project which would result in even a partial reduction in the recharge of an ecologically important aquifer could not be approved under Article 6(3). Of course, it might also plausibly be argued, using the modes of creative legislative interpretation outlined above, that consideration of the risk of such cumulative impacts on relevant groundwater should also take account of climate change risks.

Linguistic analysis

Remarkably, in addition to the teleological interpretation employed by the court, the advocate general took the time to consider 'the differing language versions of Article 6(3)', including those in English, French, Italian, German and Dutch, concluding that

> Notwithstanding these linguistic differences, it seems to me that the same point is in issue. It is the *essential unity* of the site that is relevant. To put it another way, the notion of 'integrity' must be understood as referring to the continued *wholeness* and *soundness* of the constitutive characteristics of the site concerned. (Opinion of Advocate General Sharpston, 2012, para. 54, emphasis added)

Once the advocate general's reasoning came to focus on such qualities as unity, wholeness and soundness, it was almost inevitable that she would determine that permanent loss of any portion, however insignificant, of a key ecological feature must contravene the requirement to maintain the site's integrity. Though the court did not expressly endorse the advocate general's linguistic reasoning, it appeared to do so implicitly by referring approvingly to the relevant paragraph of her opinion, and also by identifying as the key consideration 'the lasting and irreparable loss of the whole *or part of* a priority natural habitat type whose conservation was the objective that justified the designation of the site concerned' (CJEU, 2013, para. 46), without any words of qualification regarding *de minimis* loss of such habitat type or of one of its constitutive characteristics, such as connected groundwater. As water resources, including relevant groundwater, would often amount to one of the most ecologically significant constitutive characteristics or features of a protected ecosystem, this understanding of the concept of integrity would tolerate almost no diminution or deterioration of the water resources contained therein.

Ecological integrity, groundwater resources and climate change

The adoption by the CJEU of a very strict and uncompromising understanding of the concept of ecological 'integrity' under Article 6(3) of the Habitats Directive (McIntyre, 2013) indicates a liberal approach to legislative interpretation relying on the precautionary principle to justify a very stringent standard of environmental protection. This approach concept of ecological 'integrity', which is so central to the EU regime of habitats and species protection, maximizes the potential role of EU nature conservation law in the protection of groundwater resources which have an ecological function for Natura 2000 sites, a function likely in many cases to grow in importance and vulnerability due to increased climate variability.

The precautionary reasoning employed by the CJEU in interpreting the concept of ecological 'integrity', which is only applicable in the specific context of scientific uncertainty, appears highly relevant to ecologically significant groundwater, which typically lacks comprehensive hydrological data and a clear scientific understanding of its ecological role. This uncertainty is greatly exacerbated by emerging concerns over the possible impacts of climate change on groundwater resources, due both to the direct effects of climatic variability and a modified global hydrological cycle, and the indirect effects of human responses to climate change effects, such as extended and more frequent droughts (UNESCO-IHP, 2015). Thus, ecologically important groundwater will increasingly warrant precautionary protection under EU law. As sustainably managed groundwater tends to provide a range of ecosystem services in addition to supporting protected ecosystems, EU nature conservation law is likely to operate also to safeguard groundwater resources for human needs in the face of climate change, even though this is not its primary objective.

It would unquestionably be preferable to have a purpose-built, integrated EU legislative framework for the management of groundwater resources – one that could facilitate the conjunctive management and optimal use of surface water and groundwater resources, as well as the integrative management and protection of water and biodiversity, whilst also catering to human needs – rather than today's rather fragmented EU legislative framework. However, in the absence of such an ideal legislative landscape, EU nature conservation law appears to offer the greatest prospect for groundwater protection, at least regarding those groundwater resources which play an ecologically significant role for protected Natura 2000 habitats.

Conclusion

The very strict approach adopted by the CJEU to maintenance of the 'integrity' of Natura 2000 sites, involving robust protection of the key ecological characteristics and features of such sites, represents a significant opportunity for effective protection of related groundwater resources, which receive only limited protection elsewhere in EU law. The need to safeguard such groundwater becomes ever more urgent with the risk of aggravated stress posed by the impacts of climate change. In helping sustain highly valued ecosystems, such groundwater can also help safeguard related ecosystem services, on which vital human needs often depend. The European Court's approach to EU nature conservation law may be regarded as an informal (and possibly inadvertent) step towards the integration of climate change adaptation measures into existing regulatory frameworks.

Disclosure statement

No potential conflict of interest was reported by the author.

References

Atiyah, P. S. (1988). Judicial-legislative relations in England. In R. A. Katzmann (Ed.), *Judges and legislators: Towards institutional comity* (pp. 129). Washington, DC: Brookings Institution.

Bell, J. (1983). *Policy arguments in judicial decisions*. Oxford: Clarendon.

CJEU. (2013). Case C-258/11 *Sweetman v. An Bord Pleanála*, Judgment of the Court, 11 April 2013 and Opinion of Advocate General Sharpston, 22 November 2012. Retrieved from http://eur-lex.europa.eu/legal-content/EN/TXT/PDF/?uri=CELEX:62011CJ0258&from=EN

ECJ. (1991). Case 57/89 *Commission v. Germany (Leybucht Dykes)* [1991] ECR I-883. Retrieved from http://eur-lex.europa.eu/legal-content/EN/TXT/PDF/?uri=CELEX:61989CJ0057&from=EN

ECJ. (2004). Case C-127/02 *Waddenvereniging and Vogelbeschermingsvereniging* (*Waddenzee*) [2004] ECR I-7405. Retrieved from http://curia.europa.eu/juris/showPdf.jsf;jsessionid=9ea7d0f130d53f5b0b6f463a470db4be947617f3a7f7.e34KaxiLc3eQc40LaxqMbN4PaN0Ke0?text=&docid=49452&pageIndex=0&doclang=EN&mode=lst&dir=&occ=first&part=1&cid=52132

ECJ. (2006). Case C-239/04 *Commission v. Portugal*, [2006] ECR I-10183. Retrieved from http://curia.europa.eu/juris/showPdf.jsf?text=&docid=63931&pageIndex=0&doclang=en&mode=lst&dir=&occ=first&part=1&cid=53013

ECJ. (2007). Case C-304/05, *Commission v. Italy*, [2007] ECR I-7495. Retrieved from http://curia.europa.eu/juris/showPdf.jsf?text=&docid=62977&pageIndex=0&doclang=en&mode=lst&dir=&occ=first&part=1&cid=53413

ECJ. (2012). Case C-43/10, *Nomarchiaki Aftodioikisi Aitoloakarnanias*, Judgment of the Court (11 September 2012). Retrieved from http://curia.europa.eu/juris/document/document.jsf?text=&docid=126642&pageIndex=0&doclang=EN&mode=lst&dir=&occ=first&part=1&cid=53745

EU Commission. (2000). *Communication from the commission on the precautionary principle, COM(2000) 1*. Luxembourg: Author.

EU Commission. (2002). *Assessment of plans and projects affecting Natura 2000 sites: Methodological guidance on the provisions of Article 6 (3)and (4) of the Habitats Directive 92/43/EEC*. Luxembourg: Author.

EU Commission. (2007). *Addressing the challenge of water scarcity and droughts in the European Union* (COM (2007)414 final (18 July 2007)).

EU Commission. (2012a). *Report on the revision of EU policy to combat water scarcity and drought* (COM, (2012)672 final (14 November 2012)).

EU Commission. (2012b). *Blueprint to safeguard Europe's water resources* (COM (2012)673 final (14 November 2012)).

Fennelly, N. (1996). Legal Interpretation at the European Court of Justice. *Fordham International Law Journal, 20*(3), 656.

Food and Agriculture Organisation (FAO). (2015). *Shared global vision for groundwater governance 2030*. Rome: Author.

Freestone, D. (1994). The road from Rio: International environmental law after the earth summit. *Journal of Environmental Law, 6*, 193–218. doi:10.1093/jel/6.2.193

Jans, J. H. (2010). Stop the integration principle? *Fordham International Law Journal, 33*, 1533.

Jans, J. H., & Vedder, H. B. (2008). *European environmental law* (3rd ed.). Groningen: Europa Press.

Jones, G. (2012). Adverse effects on the integrity of a European site: Some unanswered questions. In G. Jones (Ed.), *The habitats directive: A developer's obstacle course?* (pp. 151). Oxford: Hart Publishing.

Maduro, M. P. (2007). Interpreting European law: Judicial adjudication in a context of constitutional pluralism. *European Journal of Legal Studies, 1*(2), 5.

McIntyre, O. (2012). The integration challenge: Integrating environmental requirements into other policies. In S. Kingston (Ed.), *European perspectives on environmental law and governance* (pp. 125–143). Abingdon: Routledge.

McIntyre, O. (2013). The appropriate assessment process and the concept of ecological "integrity" in EU nature conservation law. *Environmental Liability, 21*(6), 203.

McIntyre, O. (2014). 'The protection of freshwater ecosystems revisited: Towards a common understanding of the "ecosystems approach" to the protection of transboundary water resources under international law. *Review of European, Comparative & International Environmental Law, 23*(1), 88–95. doi:10.1111/reel.2014.23.issue-1

McIntyre, O., & Mosedale, T. (1997). The precautionary principle as a norm of customary international law. *Journal of Environmental Law, 9*, 221–241. doi:10.1093/jel/9.2.221

Millennium Ecosystem Assessment. (2005). *Ecosystems and human well-being: Synthesis.* Washington, DC: Island Press.

Opdam, P. F. M., Broekmeyer, M. E. A., & Kistenkas, F. H. (2009). Identifying uncertainties in judging the significance of human impacts on Natura 2000 sites. *Environmental Science & Policy, 12*(7), 912–921. doi:10.1016/j.envsci.2009.04.006

Opinion of Advocate General Sharpston. (2012, November 22). Case C-258/11. *Sweetman and Others v. An Bord Pleanála.* Retrieved from http://eur-lex.europa.eu/legal-content/EN/TXT/PDF/?uri=CELEX:62011CC0258&from=EN

Organization for Economic Cooperation and Development (OECD). (2015). *Drying wells, rising stakes: Towards sustainable agricultural use.* Paris: OECD Publishing.

Scott, J. (1998). *EC environmental law.* London: Longman.

Scott, P. (2012). Appropriate assessment: A paper Tiger? In G. Jones (Ed.), *The habitats directive: A developer's obstacle course?* (pp. 103). Oxford: Hart Publishing.

Simmons, G. (2010). Habitats directive and appropriate assessment. *Irish Planning and Environmental Law Journal, 17,* 4.

Söderman, T. (2009). Natura 2000 appropriate assessment: Shortcomings and improvements in Finnish practice. *Environmental Impact Assessment Review, 29*(2), 79–86. doi:10.1016/j.eiar.2008.04.001

Therivel, R. (2009). Appropriate assessment of plans in England. *Environmental Impact Assessment Review, 29*(4), 261–272. doi:10.1016/j.eiar.2009.01.001

UNESCO-IHP. (2012). *Climate change effects on groundwater resources: A global synthesis of findings and recommendations.* Abingdon: CRC Press /Taylor & Francis.

UNESCO-IHP. (2015). *Groundwater and climate change: Mitigating the global groundwater crisis and adapting to climate change.* Paris: UNESCO.

Groundwater use in North Africa as a cautionary tale for climate change adaptation

Marcel Kuper, Hichem Amichi and Pierre-Louis Mayaux

ABSTRACT

The recent history of groundwater use in North Africa provides a cautionary tale for climate change adaptation. Even though the short-term threats of groundwater overexploitation are clear, and territorially bounded, and involve comparatively few players, in recent decades, agricultural intensification has consistently increased pressure on the available resources. Groundwater has been governed through a dynamic interplay between formal rules and informal practices that focused more on maintaining fragile socio-political compromises than on ensuring environmental sustainability. If it is to be effective, climate change adaptation will need to muster the sort of political legitimacy that sustainable groundwater management is currently lacking.

Introduction: groundwater as a buffer against climate variability?

Groundwater has been used intensively for agriculture and drinking water in a considerable number of (semi-)arid regions over the past decades. It has played the role of a buffer, enabling both the state and farmers to ignore to some extent climate variability and postpone hard allocation decisions and water demand management. The 'strategic importance of groundwater for global water and food security will probably intensify under climate change as more frequent and intense climate extremes (droughts and floods) increase variability in precipitation, soil moisture and surface water' (Taylor et al., 2013). However, the intensive use of groundwater in recent times means that this buffer capacity will be less available in the future. North Africa was identified as one of the regions in the world with the highest vulnerability to decreased groundwater resources (Döll, 2009). It now appears that at the current rate of withdrawal, groundwater cannot provide a long-term response to heightened climate variability in the region.

We argue that the history of groundwater use can provide valuable lessons for assessing the potential for climate change adaptation in the future, as climate change is 'yet another (but perhaps unique) agrarian perturbation' (Birkenholtz, 2014). First, the policy prescriptions are broadly similar, as climate change policies 'must include prudent management of groundwater as a renewable, but slow-feedback resource'

(Green et al., 2011). For instance, in Morocco's 2014 national strategy for sustainable development (Royaume du Maroc, 2014), groundwater governance is mentioned as a way to improve the resilience of the agricultural sector to the impacts of climate change. Second, many features that are prevalent in water policies are also valid for climate change policies more generally (Lascoumes, 2012). For instance, the territorial scales of administrative jurisdictions and natural phenomena are often highly disjointed; costs tend to be concentrated, whereas benefits are highly diffuse; scientific uncertainties loom large; multiple actors are involved; and enforcement is challenging. Third, sustainability should actually be easier to achieve for groundwater use than for climate change adaptation. In the first case, threats are clearer and more immediate, in particular in North Africa, where local actors have witnessed the consequences of intensive groundwater use, such as rapidly declining water tables and saline intrusion (Kuper et al., 2016). Broadly speaking, local impacts can be assessed more easily, and collective action can take place accordingly with relatively fewer players. By contrast, climate change impacts are hard to measure at a local level, and it will be more difficult to identify and involve the concerned actors.

In North Africa, a private informal groundwater economy emerged in the 1980s with the rapid extension of irrigated agriculture. Groundwater quickly became a resource for profitable irrigated agriculture and of importance politically as an 'enabler of important rural socio-economic transition' (Allan, 2007). The availability of pumped groundwater 24 hours per day, 7 days a week provided a sense of abundance, at least in the short term. Encouraged by ambitious agricultural policies, farmers increasingly sought to extend the irrigated area and intensify their farming systems, often leading to greater water use at the plot, farm and regional levels and declining groundwater resources (Kuper et al., 2016).

In the past, the social and environmental sustainability of groundwater use was addressed through formal state-sanctioned rules; the governance problem was to reinforce and implement those rules. In the face of declining water tables, since the 1990s, policies inspired by the Integrated Water Resources Management (IWRM) paradigm (e.g. Rosegrant, Cai, & Cline, 2002) have gained salience. IWRM is understood as a process which promotes 'the coordinated development and management of water, land and related resources, in order to maximize the resultant economic and social welfare in an equitable manner without compromising the sustainability of vital ecosystems' (GWP, 2000). The prescriptions offered by donors and policy makers stress the 'participatory planning and implementation process, based on sound science, that brings stakeholders together to determine how to meet society's long-term needs for water' (USAID, 2007). However, formal groundwater regulation in the Middle East and North Africa (MENA) was 'found to have minimal influence over water use policy and decision-making' (Zeitoun, Allan, Al Aulaqi, Jabarin, & Laamrani, 2012). This implies that, overwhelmed by the pace of socio-technical changes, and weakened by the effects of structural adjustment programmes, the states were failing to keep groundwater extraction through informal practices of farmers within proper bounds. This view, however, is only partially true, as current unsustainable practices reflect the mindset of both the region's water users and its governments, which is focused on 'increasing agricultural production to meet national food needs and improve irrigators' incomes' (Allan, 2007).

The argument of this article is then twofold. First, we contend that groundwater use happens in the 'interplay of formal and informal institutions, in the complex social arenas within which people actually live' (Cousins, 1997). This means that we focus neither on formal regulations, nor on informal practices of groundwater use, but rather on the multiple interactions between the two, as proposed by Cleaver (2012). Second, we stress that these mutual adjustments focus more on preserving social and political compromises than on tackling the social and environmental sustainability of groundwater use. The recent history of groundwater use in North Africa thus provides a cautionary tale for the potential for future climate change adaptation. The lessons that can be drawn from several decades of groundwater use have not been encouraging. This should lead us to recognize that no climate change adaptation will take place unless it musters the sort of political legitimacy that sustainable groundwater management currently lacks.

The analytical framework

Empirically, our interests lie in the institutions that shape groundwater use, and how these institutions can evolve to effectively mitigate overexploitation. Following Hacker, Pierson, and Thelen (2015), we believe it makes sense to think of public policies as constituting institutions insofar as they embody legally enforceable rules, creating new organizations with state-backed decision-making and/or enforcement power. Many institutionalists follow North (1990) in adopting a broad definition of institutions that includes both formal and informal arrangements. However, the work of Levitsky and Murillo (2009) shows that we are far better off if we do not conflate the two under the same heading. Only by distinguishing formal institutions, which are obligatory and subject to third-party enforcement, from informal norms, which are based on shared understandings and conventions, can we explore the connections and relationships between the two. Formal institutions are openly codified, in the sense that they are established and communicated through channels that are widely accepted as official. They can ultimately be enforced by the judiciary. By contrast, informal institutions are socially shared rules, usually unwritten, that are created, communicated, and enforced outside officially sanctioned channels. They cannot be enforced by the judiciary.

The relationship between formal and informal arrangements matters all the more because institutions are *distributional instruments* laden with power (Mahoney & Thelen, 2010; Skocpol, 1995). Therefore, any given set of rules that shapes action – whether formal or informal – will have unequal implications in terms of resource allocation. Since they represent contested settlements, institutions are always vulnerable to efforts by strategic actors to change them. Formal and informal institutions can therefore be seen as two different terrains, with differing characteristics, on which these constant renegotiations take place. The constant back-and-forth between them can be described as arena-shifting (Baumgartner & Jones, 1993). Of course, the arena of formal rule-making and enforcement, on the one hand, and informal norms, on the other, often overlap. Many actors are involved in both, and the boundary between the two is often unclear. Whatever their porousness, however, they maintain some structural differences that suggest keeping them analytically separate. The first one, as we mentioned, is that the judiciary plays no direct role in enforcing informal arrangements.

Beyond that, formal rules are often guided by concerns of appropriateness, especially the state's desire to signal that it can follow international 'best practices'. By contrast, informal norms arise more systematically out of considerations of practicality, for example by complementing, substituting or competing with ineffective formal regulations (Helmke & Levitsky, 2004). Lauth (2000) points out another difference when he argues that because informal rules 'do not possess a center which directs and coordinates their actions', informal change is likely to be more decentralized, lengthy and uneven than formal change.

The process of shifting between the formal arena and the informal one can run both ways. On the one hand, actors (either farmers or state agents) dissatisfied by the formal rules but lacking the power to change them can engage in alternative informal institution-building. On the other hand, actors dissatisfied by the informal procedures (again, either farmers or state agents) may set out to devise new formal rules on top of them. In both cases, actors engaged in this sort of strategy can best be described as *subversives* (Mahoney & Thelen, p. 25–26). These dynamics matter, because they mean that the adoption of formal, environment-friendly rules will not have straightforward effects, but only indirect ones, and the rules might be rearranged and distorted.

Building on these sets of assumptions, this article will follow the course of our research by first exploring informal arrangements around groundwater use. Most researchers, including us, begin their investigations by noting that informality is prevalent in agricultural groundwater economies (Shah, 2009). Only gradually do they take notice of the way informal arrangements are themselves shaped by formal rules, and also that what is informal today may be gradually formalized.

The case studies

This article is based on two case studies: the lower Cheliff plain (Algeria) and the Saiss plain (Morocco). In both countries, groundwater use has become an important feature of irrigated agriculture. In Algeria, 88% of the irrigated area relies on groundwater. In Morocco, better endowed with surface water, this figure is 42% (Kuper et al., 2016). Both study areas were historically shaped by profound agrarian changes and illustrate the rapid transitions to intensive agriculture as well as the increased stress on groundwater resources in North Africa. In both study areas, water tables are declining, sometimes by more than a metre per year.

Historically in both areas, French settlers expropriated the land during the colonial period (Morocco 1912–56, Algeria 1832–1962), and their large-scale farms were converted after Independence (1956 in Morocco and 1962 in Algeria) to state farms. Following the period of structural adjustments, these state farms were then handed over either to the private sector or to the former workers of these farms, who were given land-use rights while remaining under some sort of state control. This was the case in Morocco, where part of the public land was granted to private companies in 1973 to continue developing a modern agriculture. The other part of the public lands in Morocco was granted to the former workers of these large-scale farms or to landless or small peasants organized into cooperatives of agrarian reform, and referred to as land-reform beneficiaries or 'assignees'. In Algeria, on the other hand, all public lands were granted to the former workers of the state farms in the framework of the 1987 land reform. More recently, private investors have progressively entered these public lands and taken over much of the land of the former workers, either by informally renting the land (Algeria and Morocco) or by buying it (only in Morocco).

The first study area is the 3000 ha irrigation scheme of Ouarizane (Algeria), which is part of the 29,000 ha Lower Chelif large-scale irrigation system, of which only 15,800 ha was irrigated in 2013, due to the lack of surface water. This irrigation scheme is representative of the 11 large-scale public irrigation schemes in northern Algeria, whose surface areas range from 11,000 to 24,000 ha. The other 18 irrigation schemes of northern Algeria are smaller and range from 2000 to 10,000 ha. The Lower Chelif irrigation scheme was historically irrigated through surface water in a semi-arid context (< 300 mm/y of rain). Due to droughts in the early 1990s, priority was given to providing coastal cities with drinking water. Most farmers remained without surface water access for several years and were inclined to turn to groundwater, which accelerated with the arrival of lessees after the 1987 land reform.

The second study area is a 4300 ha area in the Saiss plain, which covers a total of 220,000 ha (Morocco). Following droughts in the early 1980s, farmers sought access to groundwater through wells in the phreatic aquifer and tube-wells in the phreatic and confined aquifers. Rain-fed cropping systems were gradually converted to year-round irrigated agriculture (fruit trees, horticulture, forage). Almost 80% of the study area has now access to groundwater (Ameur, Amichi, Kuper, & Hammani, 2017). Disappointed with the agricultural performance of the land-reform cooperatives, the state privatized the land in 2007, which could be bought for a modest price by the assignees. With little means for agricultural intensification, however, these assignees then sold almost 40% of their land to investors, and rented out an additional 20%. Most investors and lessees immediately installed tube-wells, and groundwater use doubled in 2005–14 (Kuper et al., 2016). At the same time, a large number of the shallow wells of the assignees ran dry due to overpumping in the area.

Results

Informal practices to obtain access to resources

The groundwater economy around the world is generally considered to exist 'entirely within the private and informal sectors, with no, or very limited, regulation' (Shah, Deb Roy, Qureshi, & Wang, 2003). In North Africa, the groundwater economy often remained seemingly invisible for policy makers and developed around a wide range of informal practices of farmers to obtain access to different resources, including land, water and credit.

First, many farmers drew water from unauthorized tube-wells. The vast majority of tube-wells in North Africa are not registered, and when they are registered there is no control on the volumes that are pumped (Kuper et al., 2016). On the state-owned land in Ouarizane (Algeria), the control on the installation of tube-wells was stricter than on private land, due to the presence of state agents. Drilling of tube-wells was difficult, as the assignees had not yet accumulated any capital, but also because the state strictly forbade tube-wells inside what was formally a surface irrigation scheme. This measure was promulgated at a national scale, to deal with the overexploitation of groundwater in some irrigation schemes, in which the process of overexploitation was advanced. However, it was applied to all the irrigation schemes in Algeria without taking into account the local specificities and especially without any precautions on the impact it could have on poor farmers. It was only at the end of the 1980s that the situation changed in Ouarizane due to the arrival of lessees. Lessees were attracted by the productive potential of the public lands that were opening up informally to them, due to the dismantling of the state farms and the failure of assignees to develop the land. They rented the land from assignees through informal arrangements. The lessees had financial capital for intensive agriculture, including the drilling of (illegal) tube-wells. They also possessed the social network to drill covertly. They knew the drillers, and knew how to ensure the discretion of neighbours (Amichi et al., 2012). Likewise, in the Mitidja plain (Algeria), Imache, Hartani, Bouarfa, and Kuper (2010) estimated that 23% of the pumped groundwater was obtained through informal transactions to irrigate horticultural crops grown by lessees whose presence was not officially acknowledged. In the other study of the Saiss (Morocco), about 50% of the assignees managed to install shallow wells. Only when the state subsidized these wells would farmers register them. Following the arrival of private investors in 2007 and the implementation of the 2008 Green Morocco Plan, these investors were keen to register their tube-wells to obtain access to agricultural subsidies, as an authorized water source was obligatory (Fofack, Kuper, & Petit, 2015). More recently, when authorization was no longer required, interest by farmers in obtaining formal authorization declined.

Second, in North Africa, access to land is often obtained through informal arrangements (Ammar Boudjellal et al., 2011). This makes it even more difficult to follow up on groundwater use, as a considerable number of farmers are not registered. Vegetable producers are some of the main users of groundwater in North Africa, as they are in other Mediterranean countries, such as Spain. In the case of horticulture, farmers are often mobile and rent the land on short informal contracts. This flexibility can be explained by the very intensive and unsustainable agricultural practices of farmers, who change plots regularly due to problems of soil fertility and diseases. These farmers are often young and depend on these contracts for access to land. On the other hand, landowners often rent out the land or engage in sharecropping to deal with cyclical constraints (lack of capital, unavailability of family labour) in order not to leave the land fallow (Amichi et al., 2012). Informal contracts typically cover 30–40% of the land in an irrigation system, and enable the pooling of all the necessary resources to farm, including land, water, capital, labour and know-how (Ammar Boudjellal et al., 2011).

Third, farmers rarely use the formal credit system, although access to capital has become increasingly important due to the more intensive farming systems. Farmers obtain informal access to credit through sales of standing crops before the harvest, lease and sharecropping arrangements as mentioned above, suppliers' credit, and lending between individuals (Daoudi & Wampfler, 2010). Related to this, farmers with informal access to land or water generally do not qualify for public subsidies, which generally go to well-established farms. In many cases, farming is also exempted from taxes. A substantial number of farms are thus not directly connected to the financial mechanisms of official agricultural policies, and thus less exposed to public incentives, for example for water-saving. On the other hand, there are many generous subsidies for established farmers related to the planting of orchards, irrigation equipment, the clearing of new agricultural land, etc., encouraging farmers to intensify agricultural production, often based on intensive groundwater use.

Fourth, the knowledge and know-how required to conduct irrigated agriculture is often obtained through informal advice from knowledge intermediaries, such as input sellers. A host of formal and informal intermediaries provide 'information, knowledge, advice, funding' and act as knowledge brokers (Poncet, Kuper, & Chiche, 2010). Farmers' practices are thus generally embedded in a larger 'grey' groundwater economy that responds better to their needs than the official services (Poncet et al., 2010). These services involve sellers of agricultural and irrigation equipment and inputs, often also providing credit and advice, mechanics handling the daily repair of machinery (pumps, engines, drip irrigation systems, agricultural machinery), and agricultural service providers (e.g. for ploughing). This makes the dialogue between state services and farmers more complicated, for example on the issue of groundwater use.

These informal practices show the scramble of farmers to obtain access to productive resources in a context of a transition to more productive agriculture, where capital takes an increasingly important place. Rightly or wrongly, farmers have interpreted their mission as having to increase agricultural productivity, as epitomized in generous agricultural subsidies. The official restrictions on groundwater use are then seen as hindrances that can be circumvented.

Despite informal practices, the state is an active but contradictory actor in the groundwater economy

In North Africa, groundwater is now firmly associated with productive irrigated farming. By extension, the groundwater economy became progressively an important and formally recognized part of what remained a national priority for North Africa's political economies (Allan, 2007). At the same time, increased pressure on (ground)water resources and the lively international debates on IWRM in the 1990s encouraged the design of new water policies. This explains why the state became an active but not always very visible actor in the groundwater economy, thus influencing the ever-changing and adaptive informal practices of groundwater use. State intervention does not obey a monolithic, uniform logic but reflects ongoing debates and balance of forces between protagonists of agricultural development, actors concerned with preventing social unrest after the Arab Spring, and the river basin agencies keen to protect water resources (Fofack et al., 2015). These three different ambitions can be detected in public policies and generally materialize in public measures such as authorizations (who has a legitimate right to use groundwater?), incentives (how are certain actors encouraged/discouraged to use groundwater, and by whom?) and sanctions (how are these rules enforced, and by whom?).

First, irrigated agriculture has been 'disproportionately prominent in national water allocation policy discourse' in MENA (Allan, 2007). Recent ambitious agricultural policies (e.g. the 2009 agricultural and rural renewal policy in Algeria, the 2008 Green Morocco Plan) promote modern 'excessively intensive' agricultural models, putting more pressure on water resources (Akesbi, 2014). Groundwater tallies perfectly with such agricultural productivity ambitions. It is often individually accessed and readily available when required, in contrast to shared surface water resources that are distributed to farmers through water turns. By promoting intensive forms of agriculture, often based on groundwater, the political economies consider groundwater as a consenting provider of water resources for as long as it lasts. The 'hydraulic mission' of the state thus continues, as suggested by Allan (2002), but in different, and often more indirect ways, for example through subsidies for fruit trees and irrigation equipment.

Second, international discourse on IWRM gained importance in the Mediterranean region in the 1990s and coincided with increasing awareness of the limits of existing water resources (Margat & Vallée, 1999). This was often attributed in the press to droughts, although hydrologists rather pinpointed the importance of the greater surface area under irrigation or groundwater use for the ever-increasing demand for irrigation and drinking water (Leduc, Pulido-Bosch, & Remini, 2017). These debates inspired recent water laws and strategies in North Africa, leading to institutional reforms (e.g. the creation of river basin agencies) and to a series of measures promoting water demand management and the rational use of water, for example the water-saving programmes involving the promotion of new irrigation technology, such as drip irrigation (Benouniche, Kuper, Hammani, & Boesveld, 2014). These water policies provided the legal framework to regulate groundwater use, including the principles of 'polluter pays' or 'user pays', the requirement of obtaining authorization for groundwater use, and the definition of acceptable levels of groundwater withdrawals. However, or because of this, only a minority of farmers registered their pumping device (Fofack et al., 2015; Imache et al., 2010).

Third, in the wake of the 2011 Arab Spring, it became more difficult to enforce unpopular restrictions on groundwater use, in particular the official authorizations for the implementation of private tube-wells. Groundwater had become a safety relief valve for large numbers of farmers, and it was difficult to regulate this valve. For instance, in Tunisia a remarkable increase in the number of tube-wells was observed recently (Dugué et al., 2014).

In the field, farmers involved in the groundwater economy interact with representatives of different state services, expressing different views on groundwater use, often inspired by the three ambitions presented above. On the one hand, the agricultural services encouraged access to groundwater following severe droughts. Especially in the 1980s and 1990s, the state provided access to groundwater through public tube-well schemes or subsidized individual tube-wells. Later on, groundwater use was mainly encouraged through indirect measures. In the Saiss, for instance, state farms were transferred to private investors, who were encouraged to intensify agricultural production, requiring massive access to groundwater. Especially following the 2008 Green Morocco Plan, farmers had frequent interactions with the agricultural services for subsidies on energy, drip irrigation and fruit trees, which can be considered indirect stimuli for groundwater use by encouraging the intensification of irrigated agriculture (Kuper, Ameur, & Hammani, 2017). The different states also made considerable efforts to provide basic infrastructure in rural areas, which facilitated the deployment of the groundwater economy. The construction of roads connected farmers to markets, and the electrification of rural areas enabled the spread of powerful pumping devices.

On the other hand, farmers were increasingly confronted by legislation requiring them to obtain official authorizations for private tube-wells. In some cases, registration was enforced or encouraged, for example when registration was required to obtain agricultural subsidies. However, even when tube-wells were registered, the state had no knowledge of the volume of water withdrawal. This does not mean that local authorities are not aware of these pumping devices. Interestingly, farmers often interacted more closely with local authorities, who were looking for social peace and viewed as more understanding of the plight of farmers, than with the unknown water authorities. Fofack et al. (2015) showed that farmers in the Saiss declared tube-wells to local authorities to be – from their viewpoint – in good standing with the law, rather than obtaining the cumbersome and expensive official permit from the river basin agency. In other words, groundwater users are generally aware and make use of the contrasting discourses of the various state services. It should be added that the state remained a powerful player in the irrigation sector in North Africa due to the political importance of irrigation in the building of the nation. Ultimately, groundwater users who are in difficulty due to declining water resources sooner or later call on the state to find solutions. Private actors look for public protection, and the state was not only legally responsible for groundwater resources but was also held morally responsible for finding solutions (Kuper et al., 2016).

Informal practices supported through formal procedures and vice versa

The presence of the state in what appeared to be an informal groundwater economy shows that informal practices were in part structured by formal mechanisms, whereby farmers 'meet' a diversity of state agents and are thus confronted with the often contradictory directives of the state. The question is then how informal practices, in turn, influence public action. Confronted with the apparent contradiction of inapplicable official rules and field realities, the different actors not only adjust to official rules (Lees, 1986) but also adjust the rules. We will illustrate this through two cases.

In the Saiss, state-owned land cultivated by land-reform beneficiaries was privatized from 2007 onwards. To attract investors into agriculture, the agricultural services provided substantial subsidies for land clearing, drip irrigation and fruit trees. One of the main problems investors faced was the obligation in the subsidy procedure to obtain formal access to water. Forced to legalize their tube-well if they wanted to obtain subsidies, 3000 farmers submitted applications in 2008, their number growing to 7000 in 2013. However, the registration process required long, complicated interactions with the River Basin Agency. To overcome this bureaucratic hurdle, many investors found a way around, to ensure that their tube-well was registered quickly. For example, a 2009 decree organized a fast-track procedure to legalize older wells dating to before the 1995 water law. All that was required then was a declaration, by the farmer, that the same surface of land was already irrigated before. In other cases, farmers would get in touch with the more comprehensive local agricultural services or the local authorities in an attempt to circumvent the agency. Faced with the evidence that the registration process did not act as a filter but was in fact thoroughly undermined, the Ministry of Agriculture opted to formally admit this futility. It was therefore no surprise that from 2013 onwards, an investor no longer needed official authorization from the river basin agency to obtain agricultural subsidies.

Agricultural dynamics in Ouarizane have been assured since the water crisis of the 1990s by two processes that fall completely outside the law: access to groundwater through tube-wells and the arrival of lessees who installed them. The relations between the assignees, who were in financial difficulty and did not have access to groundwater, and the lessees, who brought in capital and agricultural know-how, were established through a multiplicity of informal contracts. Informal practices were undertaken in the early 1990s with great discretion. The actors did not know how the state would react at the local level, and informal arrangements were implemented only between actors who knew each other well (neighbours, family members). Today these practices have become so common that nobody asks whether they are authorized by the state, particularly regarding the leasing of public lands.

The ambition of lessees to assert their presence on public lands took place in a period in which the Algerian state was preparing a major land reform, which aimed to regulate such informal practices. This reform was actively debated between different interest groups within the government: those in favour of selling the land to private actors and those who wanted to maintain the land under the control of the state. In preparation since 2000, the law was finally adopted in 2010, maintaining the land in state ownership but opening up the possibility for external investors to engage in partnerships with the assignees. This provision implicitly integrates the presence of lessees on public land and thus opens the door for their possible integration through official grants. The generalization of informal practices on public lands enabled the proponents of land liberalization in the government to acknowledge the presence of lessees as essential actors in the dynamics of irrigated agriculture in the northern plains of Algeria. Informal practices thus ended up having an impact on the formulation of the land reforms.

Discussion

Formal-informal arrangements as a way to keep environmental sustainability off the political agenda

We have shown that the complex interplay between formal institutions and informal practices governing groundwater use cannot be explained away by the supposed weakness of North African states. The main problem, in short, is not that North African bureaucracies know too little, have too little money or lack coercive power over the rural world. Rather, the 'policy of informalization' that is often privileged is a way to manage the contradictions of the state in its programmes and practices, contradictions that occur between agricultural development, the maintenance of social peace and the preservation of water resources (Amichi, Mayaux, & Bouarfa, 2015). Formal-informal arrangements make these contradictions tolerable and manageable in at least two ways.

First, informality allows the obscuring of worrying trends, reducing the sense of urgency and keeping potentially unpopular decisions off the agenda. The 'tolerant' state in North Africa often encouraged farmers to appropriate access to groundwater resources, favouring economic development and social welfare (Brochier-Puig, 2004). The limits of groundwater exploitation are made even more opaque in a context of limited knowledge of atomistic groundwater use and the complexity and invisibility of groundwater dynamics. These limits are thus seen as 'adaptable' to economic and social requirements (Brochier-Puig, 2004). In other words, formal-informal groundwater management acts as a political 'safety valve' which allows people to buy time and to postpone hard allocation choices. Informal practices thus enable formal non-decisions.

Second, formal-informal arrangements enable constant, highly flexible and individualized micro-negotiations between farmers and the state. What formal arrangements do is to put new financial, legal and human resources on offer. Informal arrangements then come about for access to these resources, as shown by the case studies. In Ouarizane, for instance, it was only after the 1987 land reform, which transformed the status of workers of state farms to assignees, with the usufruct of public lands, that informal tenure relations were put in place. Likewise, in the Saiss, informal arrangements around access to groundwater were greatly shaped by access to formal resources such as authorizations and subsidies (Fofack et al., 2015). This pattern of interactions is not limited to North Africa, as shown by Allan (2007) for MENA as a whole, and by Shah (2009) for South Asia. In all these countries, public policies worked fine as long as they 'ran along with the energy and ingenuity' of informal practices of farmers, but came 'unstuck' when they tried to regulate these practices (Shah, 2009).

However, formal-informal arrangements do not only serve to manage political contradictions. In fact, formal regulations are also poorly enforced, and completed by informal ones, insofar as they are adopted also to bolster the international legitimacy of the state. This external legitimacy is not only a way to secure more funding from financial institutions, to prevent too harsh a future treatment from them, or to attract foreign private investments. Building the narrative of a 'model' country, scrupulously adhering to international norms, is also a way to foster legitimacy with its own population (Hibou, 2009). The rapid adoption of the IWRM 'toolkit' fits this strategy. The IWRM paradigm inspired the national water laws in North Africa, prompting for instance the creation of river basin agencies. It sent a powerful signal of modernity and environment-friendliness, even though the inclusion of IWRM principles 'into water policy anticipates by some decades the politically feasible circumstances which will facilitate the adoption of the new approaches' (Allan, 2003). Interestingly, the official figures on groundwater use for irrigation in North Africa are generally below the IWRM official value of 'renewable groundwater resources', for example as reported in the FAO AquaStat database, even though most aquifers in the Mediterranean were officially declared overexploited (Margat & Vallée, 1999).

More recently, tackling climate change has also become a new norm of 'appropriateness' (March & Olsen, 1989) for North African states. National strategies are being adopted, such as the one in Morocco on sustainable development (Royaume du Maroc, 2014). While institutional frameworks are still vague and poorly fleshed out, priorities include improving flood protection plans and dealing with other extreme weather events, as well as developing better information systems and promoting technological innovations for mitigation and adaptation. Within this emerging discourse, groundwater is now explicitly seen as a strategic reserve, providing security in the face of diminishing availability of surface water. Aquifer contracts, especially, are being repackaged as a policy tool for climate change adaptation (see objective 29.2 in Royaume du Maroc, 2014), whereas the two were not explicitly linked before. Combatting groundwater overexploitation, therefore, is more and more considered a climate change policy in itself.

The risk, however, is that climate change policies will meet exactly the same fate as the IWRM package, where environmental sustainability is part and parcel of formal policy, but undone by informal practices. This is the first dimension of our cautionary tale for climate change adaptation, on the basis of our analysis of groundwater governance. The quest for international standing is in itself hardly a good driver of effective policy implementation. If North African states are more concerned to signal their virtue than to effectively tackle climate change, then it will give all the more room for formal-informal arrangements, as previously described, to reassert themselves and to undermine official goals.

Resilience to climate change there will be – but for whom?

We have so far discussed the worrying incapacity to tackle groundwater overexploitation, even though it is increasingly seen as key for climate change adaptation. Another problem is that, generally, the debate on groundwater overexploitation considers that all farmers in local communities are equally concerned by its negative effects (Ameur et al., 2017). Our analysis of groundwater governance shows that this is far from being the case. This is the second dimension of our cautionary tale for climate change adaptation. Political compromise can be at the same time subtle and flexible but also highly skewed socially. Private groundwater use may reproduce and heighten deep social inequalities (Amichi et al., 2012). In North Africa, 60% of the irrigated land now depends totally or partially on groundwater (Kuper et al., 2016). However, only about 20% of the farmers have access to this overexploited resource. For instance, the progressive exclusion of land reform beneficiaries in the Saiss from groundwater access due to informal overpumping of investors and lessees cannot be observed by the river basin agency, as it settles for an estimated total abstracted volume for the entire agricultural sector, without distinguishing different social categories of farmers (Ameur et al., 2017). Furthermore, providing a single figure for groundwater use at the national level, or even figures at the basin scale, does not provide any lever to reduce water demand, as it is not clear who is overexploiting, where, or for which use. Here we see again that state agents and their formal instruments (procedures, subsidies, maps, modes of calculation) have a deeply ambivalent relationship with informal dynamics: they shape them as much as they are overcome and blinded by them (Mitchell, 2002). In the absence of formal mechanisms to manage social disparities, the tolerance of informal practices by the state in fact enables specific social categories to continue overexploitation for productive agriculture, thereby benefitting from official agricultural policies.

Moreover, in the groundwater economy, processes of exclusion occur over long periods of time, obscuring the role of political choices in favouring productive agriculture. This corresponds to what neo-institutionalist sociologists have called 'policy drift', whereby major changes occur not because of any discernible institutional change but because unchanged institutions face a fast-changing environment (Streeck & Thelen, 2005, p. 29).

Therefore, although some keen observers have qualified the process through which these inequalities were constructed as a 'pitiless struggle to access a limited resource' (Popp, 1986), these new inequalities have so far not gained any political importance. Only in some very specific cases of rapid groundwater depletion, impacting drinking water for the population, has some social unrest emerged to question the informal practices of groundwater use for productive agriculture (Houdret, 2012).

Conclusion: forging political legitimacy for climate change adaptation

We have shown that in North African countries, formal IWRM-inspired rules on groundwater use were poorly enforced and routinely subverted by informal norms for a variety of reasons. These formal rules are often designed not so much to be implemented, at least in the short run, but to signal international conformity. Also, their application would force hard and unpopular allocation choices. Finally, they are too inflexible for the daily requirements of political micromanagement. This lack of willingness to comply is, in the classical, Weberian sense, the very definition of a lack of legitimacy (Weber, 1922). Understandably, irrigators are not willing to pay the brunt of the transition costs towards more sustainable policies. They do not want to be told to draw less water when their revenues and their business model depend on water. Conversely, political elites are understandably reluctant to confront farmers. All over North Africa, rural areas are a crucial social base for the regimes in place. They are embedded in complex networks of political exchanges with the state through subsidies, paternalism and face-to-face interactions with local state agents. These agents have a fine-grained knowledge of their territory and the socio-political climate of the day. What groundwater management has shown so far is that it is unrealistic to expect the state to use coercion to transition towards better, more sustainable water policies. North African states are unlikely, in the near future, to close wells, to refuse to authorize new ones, to scrupulously enforce water quotas for extraction, or to increase water rates; at least, not to an extent that would truly make a difference (Sowers, Vengosh, & Weinthal, 2011). Transitions to sustainable policies, whether in terms of groundwater use or climate change adaptation more generally, will therefore have to be carried out with the active support of farmers and the rural areas in general, and certainly not against their will. In other words, they will have to gain wide legitimacy.

All North African states currently spend a considerable amount on agriculture. The main issue, therefore, is less to find new public resources than to redirect existing subsidies and technical support to promote environmentally sounder farming practices. States could subsidize the adoption of low-consumption crops, or encourage crops that rely mostly on green water. They could reward individual farmers for actual water savings. They could finance public works that enable better recharge of aquifers. They could reconstitute their capacity for technical assistance to farmers, which has been severely eroded during the structural-adjustment era. In short, massive new public investments are not needed. But existing ones should be redirected. This redirection will not happen overnight, however, as it has certain preconditions. First, a new 'common sense' has to be developed, disseminated and sustained among state bureaucracies, at all levels. Large parts of state bureaucracies are still imbued with a 'high modernist' ethos that value large farms with single crops, intensification at all costs and mechanization as signs of development and modernity (Scott, 1998). These sectors should gradually come to the view that 'modernity' is not reduced to one type of farming, even if export crops can bring some much-needed revenue to state coffers. Here, changes will probably start in the agricultural universities and schools, where students could be exposed to more diverse views on what could be viewed as 'good' and 'modern' farming practices. But we are still in the early stages of this process. Second, financing change is not enough. The future transition will have to be relentlessly justified and promoted to farmers. A 'communication discourse' (Schmidt, 2013) will have to be thoroughly devised that highlights short-term and longer-term benefits, underlines the costs of the status quo, heightens the sense of urgency to act, and gives a positive, modernistic image of environment-friendly farming. Third, history has shown that, most of the time, environmental policies only spring from a synergy between the state and a militant civil society. State officials rarely override powerful interests unless they are pressured to do so by vibrant, environmentally conscious and *autonomous* social organizations. This is currently what is most sorely lacking in North African countries. Agricultural trade unions lack autonomy from the state and do not push at all for sustainable transitions; urban-based environmental NGOs are weak and mainly focused on urban quality of life; and there is no political party of any importance that puts environmental policies front and centre (Lynch, 2012). Such a civil society cannot be designed by decree and depends on wider dynamics of political opening. There is some hope, however, as North African polities have entered more unsettled times. Despite all the risks that the new era entails, there seems to be no going back when it comes to popular demands for more accountability.

Disclosure statement

No potential conflict of interest was reported by the authors.

References

Akesbi, N. (2014). *Le Maghreb face aux nouveaux enjeux mondiaux. Les investissements verts dans l'agriculture au Maroc.* Paris: IFRI. http://dev.ocppc.lnet.fr/sites/default/files/IFRI_noteifriocpnakesbi.pdf

Allan, J. A. (2002). *The Middle East water question: Hydropolitics and the global economy.* London and New York: I.B. Tauris.

Allan, J. A. (2003). Integrated water resources management is more a political than a technical challenge. *Developments in Water Science, 50*, 9–23.

Allan, J. A. (2007). Rural economic transitions: Groundwater uses in the Middle East and its environmental consequences. In M. Giordano & K. G. Villholth (Eds.), *The agricultural groundwater revolution: Opportunities and threats to development* (Vol. 3). Wallingford, UK: CABI.

Ameur, F., Amichi, H., Kuper, M., & Hammani, A. (2017). Specifying the differentiated contribution of farmers to groundwater depletion in two irrigated areas in North Africa. *Hydrogeological Journal*, on-line first. doi:10.1007/s10040-017-1569-1

Amichi, H., Bouarfa, S., Kuper, M., Ducourtieux, O., Imache, A., Fusillier, J. L., … Chehat, F. (2012). How does unequal access to groundwater contribute to marginalization of small farmers? The case of public lands in Algeria. *Irrigation and Drainage, 61*(1), 34–44. doi:10.1002/ird.1660

Amichi, H., Mayaux, P. L., & Bouarfa, S. (2015). Encourager la subversion: Recomposition de l'État et décollectivisation des terres publiques dans le Bas-Chéliff, Algérie. *Politique Africaine, 137*(1), 71–93. doi:10.3917/polaf.137.0071

Ammar Boudjellal, A., Bekkar, Y., Kuper, M., Errahj, M., Hammani, A., & Hartani, T. (2011). Analyse des arrangements informels pour l'accès à l'eau souterraine sur les périmètres irrigués de la Mitidja (Algérie) et du Tadla (Maroc). *Cahiers Agricultures, 20*(1–2), 85–91.

Baumgartner, F. R., & Jones, B. D. (1993). *Agendas and Instabilities in American Politics.* Chicago: University of Chicago Press.

Benouniche, M., Kuper, M., Hammani, A., & Boesveld, H. (2014). Making the user visible: Analysing irrigation practices and farmers' logic to explain actual drip irrigation performance. *Irrigation Science, 32*(6), 405–420. doi:10.1007/s00271-014-0438-0

Birkenholtz, T. (2014). Knowing climate change: Local social institutions and adaptation in indian groundwater irrigation. *The Professional Geographer*, *66*(3), 354–362. doi:10.1080/00330124.2013.821721

Brochier-Puig, J. (2004). Société locale et État face aux limites de la ressource eau (Nefzaoua, Sud- Ouest tunisien). In M. Picouët, M. Sghaier, D. Genin, A. Abaab, G. Henri, and M. Elloumi (Eds.), *Environnement et sociétés rurales en mutation, approches alternatives* (pp. 307–321). Paris: Éditions IRD.

Cleaver, F. (2012). *Development through bricolage: Rethinking institutions for natural resource management*. Abingdon, UK: Routledge.

Cousins, B. (1997). How do rights become real? Formal and informal institutions in South Africa's land reform. *IDS Bulletin*, *28*(4), 59–68. doi:10.1111/j.1759-5436.1997.mp28004007.x

Daoudi, A., & Wampfler, B. (2010). Le financement informel dans l'agriculture algérienne: Les principales pratiques et leurs déterminants. *Cahiers Agricultures*, *19*, 243–248.

Döll, P. (2009). Vulnerability to the impact of climate change on renewable groundwater resources: A global-scale assessment. *Environmental Research Letters*, *4*(3), 035006. doi:10.1088/1748-9326/4/3/035006

Dugué, P., Lejars, C., Ameur, F., Amichi, F., Braiki, H., Burte, J., ... Kuper, M. (2014). Recompositions des agricultures familiales au Maghreb: Une analyse comparative dans trois situations d'irrigation avec les eaux souterraines. *Revue Tiers Monde*, (4), 99–118. doi:10.3917/rtm.220.0101

Fofack, R., Kuper, M., & Petit, O. (2015). Hybridation des règles d'accès à l'eau souterraine dans le Saiss (Maroc): Entre anarchie et Léviathan? *Etudes Rurales*, *196*, 127–150.

Green, T. R., Taniguchi, M., Kooi, H., Gurdak, J. J., Allen, D. M., Hiscock, K. M., ... Aureli, A. (2011). Beneath the surface of global change: Impacts of climate change on groundwater. *Journal of Hydrology*, *405*(3), 532–560. doi:10.1016/j.jhydrol.2011.05.002

GWP (Global Water Partnership) (2000). *Integrated water resources management*. TAC Background Paper, n°4., Stockholm: Global Water Partnership

Hacker, J. S., Pierson, P., & Thelen, K. (2015). Drift and conversion: Hidden faces of institutional change. In J. Mahoney & K. Thelen (Eds.), *Advances in comparative historical analysis* (Chap. 7, pp. 180–208). Cambridge, UK: Cambridge University Press.

Helmke, G., & Levitsky, S. (2004). Informal institutions and comparative politics: A research Agenda. *Perspectives on Politics*, *2*(4), 725–740. doi:10.1017/S1537592704040472

Hibou, B. (2009). Le réformisme, grand récit politique de la Tunisie contemporaine. *Revue D'histoire Moderne Et Contemporaine*, (5), 14–39. doi:10.3917/rhmc.565.0014

Houdret, A. (2012). The water connection: Irrigation and politics in southern Morocco. *Water Alternatives*, *5*(2), 284–303.

Imache, A., Hartani, T., Bouarfa, S., & Kuper, M. (2010). *La Mitidja, 20 ans après: Réalités agricoles aux portes d'Alger*. Algiers: Editions Alpha.

Kuper, M., Ameur, F., & Hammani, A. (2017). Unraveling the enduring paradox of increased pressure on groundwater through efficient drip irrigation. In J. P. Venot, M. Kuper, & M. Z. Zwarteveen (Eds.), *Drip irrigation for agriculture: Untold stories of efficiency, innovation and development* (pp. 85–104). Earthscan Studies in Water Resource Management series. London: Routledge.

Kuper, M., Faysse, N., Hammani, A., Hartani, T., Marlet, S., Hamamouche, M. F., & Ameur, F. (2016). Liberation or Anarchy? The Janus nature of groundwater use on North Africa's new irrigation frontiers. In A. Jakeman, O. Barreteau, R. Hunt, J. D. Rinaudo, & A. Ross (Eds.), *Integrated groundwater management: Concepts, approaches and challenges* (pp. 583–615). Springer:Dordrecht, The Netherlands. ISBN 978-3-319-23576-9

Lascoumes, P. (2012). *Action publique et environnement*. Paris: PUF.

Lauth, J. S. (2000). Informal Institutions and Democracy. *Democratization*, *7*(4), 21–50. doi:10.1080/13510340008403683

Leduc, C., Pulido-Bosch, A., & Remini, B. (2017). Anthropization of groundwater resources in the Mediterranean region: Processes and challenges. *Hydrogeological Journal*, on-line first. doi:10.1007/s10040-017-1572-6

Lees, S. H. (1986). Coping with bureaucracy: Survival strategies in irrigated agriculture. *American Anthropologist, 88*(3), 610–622. doi:10.1525/aa.1986.88.issue-3

Levitsky, S., & Murillo, M. V. (2009). Variation in institutional strength. *Annual Review of Political Science, 12*, 115–133. doi:10.1146/annurev.polisci.11.091106.121756

Lynch, M. (2012). *The Arab uprising: The unfinished revolutions of the New Middle East.* New York: Public Affairs.

Mahoney, J., & Thelen, K. (2010). A theory of gradual institutional change. *Explaining Institutional Change: Ambiguity, Agency, and Power* (pp. 1–37). Cambridge, UK: Cambridge University Press.

March, J. G., & Olsen, J. P. (1989). *Rediscovering institutions: The organizational basis of politics.* New York: Free Press.

Margat, J., & Vallée, D. (1999). *Vision méditerranéenne sur l'eau, la population et l'environnement au XXIème siècle.* Sophia Antipolis: Plan Bleu.

Mitchell, T. (2002). *Rule of experts: Egypt, techno-politics, modernity.* Berkeley, CA: University of California Press.

North, D. (1990). *Institutions, institutional change and economic performance.* Cambridge: Cambridge University.

Poncet, J., Kuper, M., & Chiche, J. (2010). Wandering off the paths of planned innovation: The role of formal and informal intermediaries in a large scale irrigation scheme in Morocco. *Agricultural Systems, 103*(4), 171–179. doi:10.1016/j.agsy.2009.12.004

Popp, H. (1986). L'agriculture irriguée dans la vallée du Souss. Formes et conflits d'utilisation de l'eau. *Méditerranée, 59*(4), 33–47. doi:10.3406/medit.1986.2425

Rosegrant, M. W., Cai, X., & Cline, S. A. (2002). *Global water outlook to 2025.* Colombo, Sri Lanka.: Averting an impending crisis. IWMI.

Royaume du Maroc (2014). Stratégie Nationale de Développement Durable, Rapport final, 134 p. http://www.environnement.gov.ma/PDFs/SNDD-Rapport-Final-2015.pdf

Schmidt, V. A. (2013). Democracy and legitimacy in the European Union revisited: Input, output and 'throughput'. *Political Studies, 61*(1), 2–22. doi:10.1111/j.1467-9248.2012.00962.x

Scott, J. C. (1998). *Seeing like a state: How certain schemes to improve the human condition have failed.* New Haven, CT: Yale University Press.

Shah, T. (2009). *Taming the Anarchy: Groundwater Governance in South Asia.* Washington DC: Resources for the Future Press.

Shah, T., Deb Roy, A., Qureshi, A. S., & Wang, J. (2003). Sustaining Asia's groundwater boom: An overview of issues and evidence. *Natural Resources Forum, 27*(2), 130–141. doi:10.1111/narf.2003.27.issue-2

Skocpol, T. (1995). *Social policy in the United States: Future possibilities in historical perspective.* Princeton, NJ: Princeton University Press.

Sowers, J., Vengosh, A., & Weinthal, E. (2011). Climate change, water resources, and the politics of adaptation in the Middle East and North Africa. *Climatic Change, 104*(3–4), 599–627. doi:10.1007/s10584-010-9835-4

Streeck, W., & Thelen, K. (2005). Introduction: Institutional change in advanced political Economies. In W. Streeck & K. Thelen (Eds.), *Beyond continuity: Explorations in the dynamics of advanced political economies* (pp. 1–39). New York: Oxford University Press.

Taylor, R. G., Scanlon, B., Döll, P., Rodell, M., Van Beek, R., Wada, Y., ... Konikow, L. (2013). Ground water and climate change. *Nature Climate Change, 3*(4), 322–329. doi:10.1038/nclimate1744

USAID. 2007. What is integrated water resources management? www.usaid.gov/our_work/environment/water/what_is_iwrm.htm

Weber, M. (1922). *Wirtschaft und Gesellschaft (Grundriß der Sozialökonomik)* (1st ed.). Tübingen: Mohr.

Zeitoun, M., Allan, T., Al Aulaqi, N., Jabarin, A., & Laamrani, H. (2012). Water demand management in Yemen and Jordan: Addressing power and interests. *The Geographical Journal, 178*(1), 54–66. doi:10.1111/j.1475-4959.2011.00420.x

Global climate change and global groundwater law: their independent and pluralistic evolution and potential challenges

Joyeeta Gupta and Kirstin Conti

ABSTRACT
Although the climate and groundwater systems have close links, the international climate change regime and global groundwater laws have developed independently, despite being negotiated within a few years of each other. Hence this article addresses the question: Do global legal instruments on climate change and groundwater consider the geophysical links between the two systems, and how can their legal frameworks be improved? It argues that there are six geophysical links between groundwater and climate change which are presently inadequately accounted for in the legal regimes and there are four key contradictions between the two legal systems. It makes four recommendations to enhance the linkages between the systems.

Introduction

Climate change and groundwater problems are closely linked (Bates, Kundzewicz, Wu, & Palutikof, 2008; Jiménez Cisneros et al., 2014), but their international governance has evolved independently. It is then natural to ask: Do global legal instruments on climate change and groundwater consider the geophysical links between climate change and groundwater, and how can their legal frameworks be improved? To address this question, this article (a) analyzes the literature on geophysical and related aspects of climate change and groundwater; (b) undertakes a content analysis of the relevant global laws; and (c) uses legal pluralist, politics of scale and hydro-hegemony theories.

The article focuses on groundwater rather than freshwater for two reasons. First, groundwater is 97% of available freshwater (Margat & van der Gun, 2013), which itself is a small percentage of total water. The volume of groundwater recharged is thrice that of total surface water flows over the last 50 years (Gleeson, Befus, Jasechko, Luijendijk, & Cardenas, 2015), and its abstraction is rising by 3% annually (Wada, Wisser, & Bierkens, 2013). In 2010, it provided 50% of potable water (Margat & van der Gun, 2013) and 40% of irrigation needs (Döll et al., 2012; Siebert et al., 2010). Second, there is a tendency for governance to focus only on fresh surface water and incidentally on groundwater. Specifically referring to groundwater may make the 'invisible' more 'visible'.

The analysis uses legal pluralism, the politics of scale and hydro-hegemony theories. *Legal pluralism* (Von Benda Beckmann, 2001) refers to different rules emerging from (in) formal actors at varying governance levels applying to the same jurisdiction (Zips & Weilenmann, 2011). This can lead to contradictions, when multiple systems coexist, or fragmentation, when it evolves in a bottom-up manner or because top-down consensus reaches an impasse (Koskenniemi & Leino, 2002; Tamanaha, 2008). A *politics of scale* lens helps examine why states may or may not scale up an issue and its sub-parts to the global level (Gupta, 2008), while *hydro-hegemony* scholars further explain why and how powerful states use their power to control the shape of water agreements, their interpretations and their ratification (Mirumachi, 2015; Nicol & Cascão, 2011; Zeitoun & Allan, 2008).

This article first examines the physical relationship between climate change and groundwater, then their governance and their normative and political contradictions, before drawing conclusions and offering recommendations.

Physical relationship between climate change and groundwater

Climate change drives changes in the hydrological cycle and is exacerbated by how groundwater is used. Further, these relationships exist in complex self-reinforcing ways at multiple levels.

Climate change affects the hydrological cycle and its management

Climate change affects the hydrological cycle through greater evaporation, sea-level rise, melting glaciers and ice caps, changing rainfall patterns, and extreme weather events (Bates et al., 2008). The impacts on groundwater quantity and quality are uncertain because of the multiple feedback effects (Jiménez Cisneros et al., 2014) – depending on the magnitude, intensity, seasonality, frequency and location of precipitation, types of ground cover, existing soil moisture, and geological setting, combined with physical and human interference in water flows – affecting groundwater recharge. Thus, greater rainfall does not necessarily increase recharge. Sea-level rise can influence groundwater quality through saltwater intrusion (Nicholls & Cazenave, 2010). Changes in groundwater levels and recharge mechanisms can mobilize new contaminants from the (sub)surface and release them into aquifers (Green et al., 2011). Furthermore, climate change will affect hydropower, flood defences, irrigation and water supply systems (Bates et al., 2008).

Groundwater use can exacerbate climate change

Groundwater use can also exacerbate climate change. First, energy production and land-use changes emit greenhouse gases (GHGs) and use water. Firewood uses the most water, followed by hydropower, nuclear, oil, coal and lignite, geothermal, natural gas and solar, with wind energy's footprint being marginal (Mekonnen, Gerbens-Leenes, & Hoekstra, 2015). Land-use change through deforestation and draining of wetlands can simultaneously emit GHGs and affect groundwater. Second, the water sector uses energy (e.g., 20–30% in California); groundwater pumping, long-distance transfers, and desalination use mostly fossil-fuel energy (Hanak & Lund, 2012). Third, when groundwater is pumped and discharged from non-recharging aquifers, it eventually contributes to sea-level rise – possibly about 0.25 mm between 1990 and 2000, and potentially 0.87 mm by 2050 (Wada et al., 2012).

Groundwater and climate change at multiple geographic levels

The climate and hydrological systems are global, with effects that manifest or are reinforced locally; their mutual relationship is influenced by geography, geomorphology, and policies and practices at multiple governance levels. At the regional level, the Sahel, Siberia and the western US may have more groundwater recharge, while southwest Africa and southern Europe may have less (IPCC, 2007). Water stress could increase in West and South Asia (especially India), Southern and North Africa, Central America, and much of Europe. Under all climate scenarios, all regions will lose groundwater resources, except North Africa because of the great depth to groundwater there (Ranjan, Kazama, & Sawamoto, 2006).

At the national level, groundwater-dependent countries such as the US, India, Australia and several West Asian and African countries use managed aquifer recharge techniques to store excess surface water or treated wastewater. This capitalizes on groundwater's buffering capacity, increases reliability of supplies, and combats contamination (Margat & van der Gun, 2013). Small island states may be flooded by sea-level rise, while saltwater may enter groundwater in low-lying coastal areas (Treidel, Martin-Bordes, & Gurdak, 2011). At the local level, differentiating between management- and/or climate-induced impacts is important for appropriate adaptation responses (Heuvelmans, Louwyck, & Lermytte, 2011).

Inferences

Table 1 sums up the above key links between climate change and groundwater and their implications for governance.

Relationship between climate change and groundwater governance

This section examines the global co-evolution of climate change and groundwater law (Figure 1) in terms of history, architecture, goals and process.

Climate law history and integration of water issues

Climate change entered the global scientific agenda in 1979 and the global political agenda in 1989. Governance responded to a science-driven process. Within two years the United Nations Framework Convention on Climate Change (UNFCCC, 1992) was adopted; it is now universally adopted and in force. Thus, global climate policy preceded and shaped national climate policy (Bodansky, 1993; Gupta, 2014a). Global climate governance occurs under the Climate Convention, its annual Conference of the Parties, and its decisions (i.e., the climate governance regime), the most prominent of which are the Kyoto Protocol of 1997 (in force), with emission targets for developed countries for 2008–2012, the Doha Amendment of 2012 (not yet in force), with targets

Table 1. Implications of the links between climate change and groundwater for law.

Systems	Issues	Implications for law
Climate change impact on groundwater	Quantity of recharge and impact on flows	Needs to create adaptive policies accounting for climate change's influence on flows and recharge patterns; revise spatial planning to maximize recharge; use groundwater as a buffer; enhance transboundary governance arrangements
	Inundation, saltwater intrusion, and indirect impacts on water quality through impacts on quantity	Needs to link with coastal defence and agricultural policy to manage and reduce saltwater intrusion; needs to anticipate other quality impacts and take action accordingly
	Intensity and frequency of extreme weather events	Needs to link up with disaster risk reduction policy at the global level and integrate into drought and flood strategies using groundwater as a buffer
Groundwater impact on climate change	Energy use in water extraction and use	Needs to make policies to reduce the energy intensity of water uses
	Extraction leading to sea-level rise and changes in groundwater quality	Needs to ensure that policy keeps groundwater recharge in line with extraction to maintain water quality and quantity
	Water dependency of energy production	Climate and groundwater law needs to ensure water accounting in energy policy

for developed countries for 2012–2020, and the Paris Agreement of 2015 (in force), with bottom-up targets for all countries which need to be ratcheted up every five years. Desertification was left to the Convention to Combat Desertification (UNCCD, 1994), and deforestation only became part of the climate negotiation process after 2005. The

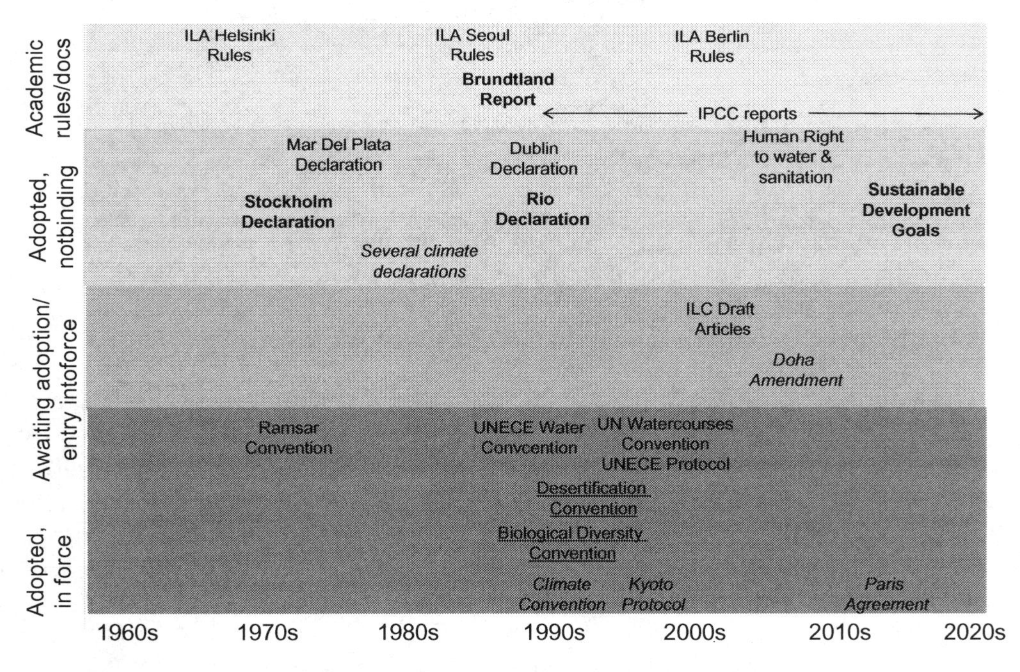

Figure 1. The evolution of climate change and groundwater-related agreements at the global level (normal text represents agreements on water; italic, climate change; underlined, environment; bold, environment and development).

climate regime includes measures on energy, but scarcely mentions water governance regimes despite the contemporaneous adoption of the regional United Nations Economic Commission for Europe (UNECE) Water Convention in 1992 and the UN Watercourses Convention (UNWC, 1997) in 1997. But since 2003, there has been heavy lobbying to include water issues more explicitly, given the clear physical linkages between the two (Gupta, 2014a).

The climate regime aims to stabilize atmospheric GHG concentrations so that they correspond to an average global temperature that is no more than 1.5–2 °C above pre-industrial levels and thereby implicitly aims to reduce the impacts on the global water system. It includes targets, policies and measures that emerge from the principles under the Climate Convention (see below). Countries prepare relevant national policies consistent with their obligations under the convention, which they report on in their National Communications. However, they focus on energy rather than water policies. Although the climate regime is perhaps one of the most centralized, especially in comparison to the energy or water governance arenas, some scholars have described it as fragmented (Van Asselt, 2014).

The climate regime is dynamic, with five bodies, annually recurring meetings, a strong secretariat and ever-evolving institutions engaging the market and funding mechanisms (Gupta, 2014a). It is actively supported by science from the Intergovernmental Panel on Climate Change (IPCC), within which the legal epistemic community plays a minor role.

Water law and integration of climate change issues

Unlike climate change, water law has developed over centuries through customary rules, religious rules institutionalized in domestic legal systems, which were then exported through conquests and colonization processes and subsequently increasingly affected by the growing number of bilateral and multilateral water agreements on transboundary waters, jurisprudence, and global discourses over communism, environmentalism and neoliberal capitalism (Caponera, 1992; Dellapenna & Gupta, 2009). There has been a continuous interplay between the various levels of water governance, but it has always been an area of heavy politics, as those who controlled water controlled power. At the global level, water law was influenced by the International Law Association's (ILA) Helsinki Rules on the Uses of the Waters of International Rivers (International Law Association [ILA], 1966) and its Seoul Rules on International Groundwaters (ILA, 1986); the Ramsar Convention (1971) on Governing Wetlands of International Importance (hard law, universally binding); the UNECE (1992) Water Convention (hard law, now global, 38 parties) and its follow-up Protocols; the UN Convention on the Non-navigational Uses of International Watercourses (UNWC, 1997 – hard law, legally binding, 35 parties); the United Nations Millennium Declaration (United Nations General Assembly [UNGA], 2000), with its water targets (soft law, but actively implemented); the ILA's Berlin Rules (ILA, 2004 – academic codification of customary law); the ILC Draft Articles on the Law of Transboundary Aquifers (ILC, 2008; UNGA, 2008), the UN General Assembly and UN Human Rights Committee Resolutions on the Human Right to Water and Sanitation (UNGA, 2010; UNHRC, 2010 – soft law, now possibly customary law); and the Sustainable

Development Goals (UNGA, 2015 – soft law) (Conti & Gupta, 2015). Most of these dealt more with surface water than with groundwater and did not deal with climate impacts. The Draft Articles appear to have reached an impasse in the General Assembly as governments do not feel they have sufficient knowledge regarding transboundary aquifers to determine their final legal form (Eckstein & Sindico, 2014) and because the International Law Commission's substantive work and its role in promoting consensus is falling short of what is needed (Stoa, 2014). Further, these texts include groundwater in their scopes to varying degrees, and the in-force UNWC does not include all types of transboundary aquifers.

The legally binding agreements focus on transboundary water governance issues and domestic water issues when they are of international importance. The human rights documents focus on domestic access to water and sanitation services. The Berlin Rules and the Sustainable Development Goals draw on human rights and other bodies of international law to go beyond transboundary issues to also address issues that fall within domestic purview. However, none of these governance texts deal with climate-proofing water governance from a mitigation or an adaptation perspective. The Sustainable Development Goals do make an effort in this direction in that they require that the indivisible Goals are dealt with in an integrated and interrelated manner. Indirectly, other international treaties also discuss various aspects of groundwater governance – such as the Desertification and Biodiversity Conventions.

The UNWC is a static, one-time agreement without operational bodies such as a secretariat. It cannot continuously amend itself based on new scientific knowledge or legal progress. Although the now-global UNECE Water Convention and its secretariat could perhaps step into the breach, this may be less legitimate since non-UNECE countries' inputs did not shape the text. In fact, water governance is undertaken by many UN and non-UN bodies (Pahl-Wostl, Gupta, & Petry, 2008) and loosely coordinated by UN-Water (Baumgartner & Pahl-Wostl, 2013). Further, water policy and law are strongly influenced by epistemic and professional communities, such as the International Law Association and the World Water Council, and development banks, who have their own investment approaches for water.

Inferences

This section has shown that international climate and water laws have been independently negotiated and have not built on each other or the substantive relationship between the issues they deal with. While climate law regulates GHG emissions and thereby implicitly the impact on the global water system, water law does not explicitly consider the energy use of water or the way water may influence GHGs. Climate change has only recently been included in groundwater law through managed aquifer recharge provisions; other issues have scarcely been taken up.

There are also major architectural differences between the regimes. While the climate regime has global scope, because the Climate Convention and the Kyoto Protocol have near-universal ratification and the Paris Agreement has entered into force, the water agreements (barring Ramsar) have been ratified by less than a quarter of UN member states. While the Climate Convention is a framework allowing for dynamic evolution within its various bodies, the water conventions have patchy coverage, and the UNWC is a static law. While climate law explicitly focuses on both inter-state and domestic responsibilities, international water law is dispersed in different regulations, focusing on transboundary responsibilities, wetlands of global importance, and meeting the human right to water and sanitation, and is indirectly influenced by treaties on desertification and biodiversity. Climate law is internally consistent across levels, since countries are implementing it, although in a common but differentiated manner which is increasingly taking on a bottom-up approach, as in the Paris Agreement. However, the global groundwater laws are inconsistent horizontally and vertically at the different levels, creating plural approaches to water governance, which have been extensively discussed in other papers (Conti & Gupta, 2014; Gupta, Hildering, & Misiedjan, 2014; Obani & Gupta, 2014).

Key discursive, normative, substantive and political challenges

Introduction

This section compares the global climate and transboundary water law regimes in terms of discursive, normative, substantive and political challenges.

Discursive, normative and substantive challenges

We first discuss the 'global' nature of the two issue areas. Climate change was framed as a global issue from the start and has evolved into a globally steered multilevel regime. But historically, water was seen as a local issue, then as a national issue, later a transboundary issue, and only more recently as a global issue. While countries appear willing to discuss transboundary aspects, neither of the treaties on transboundary sharing have been ratified by more than a quarter of the world, and those ratifying are mostly downstream or European countries. While academic and policy documents increasingly treat water as a global concern, some countries are reluctant to deal with water as part of a global-level cycle, to address the physical differences between groundwater and surface water resources, and to regulate it in accordance with some globally decided principles. However, their universal acceptance of the Ramsar, Desertification and Biodiversity Conventions implies that they need to address water issues in the context of these treaties.

If climate change is a globally steered multilevel issue, and affects water, then why did water not get as much attention within climate change? This is because while mitigation was framed as a global challenge, adaptation (which is more closely related to water) was defined as a local challenge in the Climate Convention (Bodansky, 1993). This was done to reduce rich countries' liabilities in relation to adaptation (Gupta, 2014a), and water users and the perceived impacts of use are primarily local (Alston & Whittenbury, 2011). This arguably led to water's receiving significantly less attention within the climate change regime and groundwater receiving only a small share of the attention within that.

Furthermore, while most discourses view climate change as a common concern of humanity, groundwater is usually seen as a shared resource by lawyers, as an economic good by development banks and policy makers (International Conference on Water and the Environment, 1992), as a human right by the human rights community (UNGA, 2010), as a gift from God in Islam, and as a heritage by the European Water Framework

Directive. These discursive differences underlie diverging approaches to water governance.

While the Climate Convention only explicitly mentions sovereignty in its preamble, international customary water law has been largely structured around sovereign rights. This has changed, as the UNECE Water Convention makes no mention of sovereignty and the UNWC recognizes only sovereign equality and territorial integrity of states, i.e., that downstream states have the right to receive the waters they have always received in the past. However, the Draft Articles explicitly recognize sovereignty over transboundary aquifers and aquifer systems, subject to the principles of cooperation, equitable and reasonable use, and not causing significant transboundary harm (Eckstein, 2007; Sindico, 2011; Stephan, 2011). There is thus concern that the draft rules signal a return to sovereignty (McCaffrey, 2011, 2013; McIntyre, 2011; Stoa, 2014).

The Climate Convention addresses equity through the principle of 'common but differentiated responsibilities and respective capabilities', which requires all parties to take responsibility relative to their different GHG emissions and their different capabilities in reducing them. It gives special attention to especially vulnerable countries. In contrast, the UNECE, UNWC, and the Draft Articles include the principle of equitable and reasonable use; the latter two elaborate on factors and weights to determine countries' share of water. However, the UNWC explicitly denies that any use of water has greater priority than any other. The Draft Articles do not explicitly negate priority of use. Both give special regard to vital human needs. The UNECE does not elaborate any priorities or factors.

Regarding the environment, the Climate Convention adopts the precautionary principle, which paved the way for the long-term objective under the Paris Agreement. The UNECE Convention includes the precautionary principle, best available technologies and environmental impact assessment in a transboundary context. The Watercourses Convention does not adopt the precautionary principle but has rules on protecting and preserving water ecosystems, controlling pollution, preventing the introduction of alien species, and protecting the transboundary marine environment. The Draft Articles protect groundwater recharge and discharge zones.

The Climate Convention amends the UNGA Declaration on the Right to Development into a right to and responsibility for promoting sustainable development; the Water Treaties do not refer to development. With respect to international trade, the UNFCCC explicitly allows an open international economic system. The UNWC and UNECE Convention do not do so, but clearly function within the context of global trade and investment and thus operate *de facto* within such a system.

Why are there such differences?

There are clear discursive, normative and substantive differences within and between the legal regimes. This occurs partly as a result of the historical evolutionary processes and the interactions between actors, including scientific actors, engaged in them. But it also occurs because of the politics of scale. States may scale up issues to the global level (1) to enhance problem understanding (i.e., the nature of the global system, the indirect causes of the problem, the global thresholds and impacts, the underlying discourses); (2) to enhance policy effectiveness (i.e., to determine international responsibility, protect the common good, mobilize the international community); (3) to serve domestic interests (i.e., avoid domestic measures); or (4) for strategic extra-territorial reasons (i.e., access resources or create markets, control resources, bypass an agency) – see Table 2.

However, countries may also scale down issues, (1) to enhance problem understanding (especially the link to local perceptions, factors, impacts and context); (2) to enhance policy effectiveness (i.e., to mobilize local people); (3) to serve domestic interests (i.e., to protect national security, manage without interference, avoid international liability or pressure); and/or (4) for strategic extra-territorial interests (i.e., to divide and control/include or exclude, to avoid losing control over natural resources, to bypass another country's agency) – see Table 3.

We support the arguments for taking a multiscalar approach to climate and water which call for coherent global-to-local institutionalized approaches that continuously feed each other (Gupta & Pahl-Wostl, 2013; Pahl-Wostl et al., 2008; Vörösmarty, Hoekstra, Bunn, Conway, & Gupta, 2015). However, many powerful states behave as hegemons on transboundary water issues, prioritizing their own narrowly defined national water interests in transboundary agreements (Zeitoun & Allan, 2008; Mirumachi, 2015; Nicol & Cascão, 2011; cf. Pahl-Wostl, Gupta, & Bhaduri, 2016). This, in addition to the inherent weaknesses of the conventions (Stoa, 2014), explains countries' reluctance to ratify the UNWC (Gupta, 2016) and come to agreement on the legal form of the Draft Articles, which both push for equitable sharing rather than unilateral power politics.

Do such differences matter?

We identify thus four contradictions. (1) The links between climate change and groundwater, although academically explored (see point 2, Table 1) have yet to be incorporated substantively in either regime, and both ignore each other (see point 3). (2) The two problems are framed differently in terms of the administrative level at which they should be addressed and in terms of whether they are seen as public or private goods, and affecting human rights or property rights. Maintaining a stable climate is an issue of global concern requiring multilevel action. While there is increasing recognition of the need to address groundwater as a global concern, groundwater tends (more than other freshwater) to be seen as a local, national, or transboundary, but not global, issue. While a stable climate is generally seen as a public good, maintaining the hydrological cycle or preventing excessive discharge from non-recharging aquifers that could contribute to sea-level rise is not yet defined as a public good. Both areas are also affected by the growing neoliberal capitalist approaches that allow the privatization of and trade in water resources and carbon credits (Bernasconi-Osterwalder & Brown Weiss, 2005; Klijn, Gupta, & Nijboer, 2009). (3) There are also key differences in the way sovereignty – the right to sustainable development, equity, and environmental harm – is dealt with. (4) Finally, there are significant differences in procedures and subsidiary bodies, as well as the incorporation of scientific knowledge, making it challenging to develop good links.

Still, a key question is, do such differences matter? In the context of the Anthropocene, it is increasingly important to also have a global perspective on climate change and water problems, not least because of the reasons listed in Table 2. Both a stable climate and a

Table 2. Reasons for scaling up climate change and groundwater.

	Argument	Climate	Groundwater
1	Global systemic links	A global climate system linked to the hydrological system	A global water cycle, including non-recharging aquifers, linked to the climate system
	Indirect causes/drivers	Production, consumption, deforestation, water use	Production, consumption, trade, investment, climate change
	Global impacts and thresholds	Climate change; 1.5–2 °C threshold	SDG targets/indicators; indirect targets from climate, biodiversity, transboundary wetlands (Ramsar)
	Level playing field	Through common principles, targets, policies and standards	
	Increase negotiating space	E.g., link climate to other global issues	E.g., link groundwater to other issues
	Global discourses	Climate as common concern within neoliberal, anarchic global order prioritizing sovereignty; SDGs	Water as commodity, human right/ security within a neoliberal global order prioritizing sovereignty; SDGs
2	Determine international responsibilities	Common but differentiated responsibilities and respective capabilities principle; right to sustainable development; protection of vulnerable countries	Equity principles for water sharing, including for non-recharging aquifers; no-harm principle; quality protection; data sharing
	Protect the common good	Prevent dangerous climate change; protect stable climate	Protect the functioning of the global water system; address nexus with other issues
	Mobilize international community	International climate industry, NGOs, civil society, UNFCCC; int'l financial institutions?	International water industry, traders, NGOs, civil society, UN agencies; int'l financial institutions?
3	Avoid unilateral domestic measures	E.g., through conditionalities in Doha and Paris targets	E.g., on water supply and sanitation by calling on international support
4	Access resources, create markets	E.g., emission credits through the Clean Development Mechanism / emissions trading	E.g., water through allowing land purchase/grabbing that includes groundwater rights or through public–private partnerships
	Control resources	E.g., energy policies of other countries	E.g., groundwater management by other countries
	Avoid bilateral policy influence	E.g., through transnational networks	E.g., through aid agencies

Adapted from Gupta and Pahl-Wostl (2013), Gupta (2014b).

healthy hydrological system should be viewed as global-to-local public goods for which states need to be held responsible, individually and collectively (cf. Kaul, Conceicao, Le Goulven, & Mendoza, 2003). Second, there are clear links between climate change and groundwater, and if the groundwater-related regimes do not actively explore the options for climate-proofing their policies and activities, the latter will be out of date and ineffective (Heather & Gleick, 2011). Third, there is a need to revisit the notion of sovereignty: transboundary and global challenges call for limiting absolute territorial sovereignty by requiring states to share resources, to limit transboundary harm and to equitably share related responsibilities. However, the ILC Draft Articles and their return to sovereignty may once more legitimize the politics of downscaling issues and taking a nationalistic, securitization perspective that goes against an understanding of the nature of the water and climate system. The differences in process and how groundwater is included in the scopes of the agreements makes it very difficult to build bridges between the governance frameworks for the two issue areas because there is no clear focal point in the water world except perhaps the coordination role of UN-Water.

Table 3. Reasons for down-scaling climate change and groundwater.

	Arguments	Climate change	Groundwater
1	Local system links	Downscaled models relevant for local adaptation	Aquifer or sub-aquifer level is most important
	Local driving factors	Differ in each context	
	Local impacts/thresholds	Local impacts and thresholds need identification	Aquifer or sub-aquifer needs and standards critical
	Local frames	Subsidiarity, decentralization	Subsidiarity, decentralization
2	Protect local communities	Need for local vision	Need for watershed vision, community-based orgs.
	Avoid int'l responsibilities	To avoid reducing emissions or pay for impacts elsewhere	To avoid sharing water and paying for harm caused
	Mobilize local people	Mobilize local people to reduce emissions and adapt	Mobilize local people to conserve groundwater, pollute less, and adapt
3	Protect national interests without interference	Protect oil export/use interests, national energy security	Protect 'national' water interests
	Avoid int'l liability	For causing transboundary harm	
	Avoid collective pressure	E.g., to implement targets	E.g., to change behaviour or share data
	Use of comparative advantages	E.g., in exporting products using local technologies	E.g., on managing groundwater; in trading products using groundwater
4	To divide and control or include and exclude	Disengaging from Kyoto Protocol allowed US to explore bi/multilateral relations	Disengaging from treaties allows non-parties to follow selective cooperative/ hydro-hegemonic strategies
	To bypass an agency	Allows some to bypass UNFCCC	Allows aid agencies and industry to market ideologies (e.g., privatization, cost recovery) without global consensus

Adapted from Gupta and Pahl-Wostl (2013), Gupta (2014b).

Another key point is that even though water law does not talk about the value of water, the banks and policy processes have framed water as an economic good, leading to its privatization, hoarding and trading. Groundwater law has yet to deal with whether privatization is compatible with an understanding of the hydrological system or with the notion of water as a human right. If climate change massively influences the distribution of water, then, indirectly, the rules regarding water access and ownership may further exacerbate the negative impacts of climate change on humans.

Conclusion

This article has examined the relationship between international laws on climate change and groundwater and identified four contradictions: (1) lack of links between the agreements, despite the close substantive relationship between climate change and water; (2) differences in framing the climate and water problem; (3) differences in key norms; and (4) differences in process. Such contradictions arise through historical evolution shaped by the politics of scale, the way hegemons try to shape institutions, and how institutions subsequently restrain hegemonic activity. Countries that wish to focus on national short-term development interests may decide to ignore global rules on climate and water.

We now consider how these challenges can be addressed. First, we recommend collaboration between the IPCC, the World Water Development Reports, the International Law Association and the Global Environmental Outlook to improve understanding of climate, water and related law. Such collaboration should complement the existing IPCC Report on Climate and Water (Bates et al., 2008), which does not discuss the water and climate governance regimes. Second, we recommend using the IPCC's relationship with the Climate Regime to formally integrate the water dimension into the climate change treaty process, building on the ideas in Table 1 and making barriers to sea-level rise, promoting spatial design that encourages groundwater recharge, promoting managed aquifer recharge and geothermal techniques, wet-proofing (making space for extra water without allowing the land surface to be washed away), and enhancing adaptive capacity (Van Vliet & Aerts, 2014).

Third, we recommend collaboration between the Climate Secretariat and UN-Water as the key authority on global water issues, possibly through a formal memorandum of understanding. Although the power, resources and mandate of the two bodies are not comparable, UN-Water's members represent most of the global actors on water. Fourth, there needs to be more academic and political debate on the relationship between sovereignty and transboundary-to-global public goods; on exploring the content of the right to sustainable development; on assessing and analyzing how the differential treatment of equity in the two regimes can be harmonized; and on how the precautionary principle and the idea of not causing harm to other countries can be further explored. This brings us to the economic component of sustainable development. Given that within the neoliberal capitalist context there is an ongoing process to privatize land, water and carbon, it may be necessary to begin academic debates regarding whether and under what conditions the trade, investment and private international law regimes counter the idea of a stable climate and a healthy water system as global-to-local public goods that should be available for all. Such discussions could be further developed in the context of the implementation of the interrelated, interconnected and indivisible Sustainable Development Goals.

Disclosure statement

No potential conflict of interest was reported by the authors.

References

Alston, M., & Whittenbury, K. (2011). Climate change and water policy in Australia's irrigation areas: A lost opportunity for a partnership model of governance. *Environmental Politics, 20*(6), 899–917. doi:10.1080/09644016.2011.617175

Bates, B., Kundzewicz, Z. W., Wu, S., & Palutikof, J. (2008). *Climate change and water: Technical paper VI.* Geneva: Intergovernmental Panel on Climate Change (IPCC).

Baumgartner, T., & Pahl-Wostl, C. (2013). UN–water and its role in global water governance. *Ecology and Society, 18*(3), 3. doi:10.5751/ES-05564-180303

Bernasconi-Osterwalder, N., & Brown Weiss, E. (2005). International investment rules and water: Learning from the NAFTA experience. In E. Brown Weiss, L. Boisson de Chazournes, & N. Bernasconi-Osterwalder (Eds.), *Freshwater and international economic law* (pp. 263–288). Oxford, UK: Oxford University Press.

Bodansky, D. (1993). The United Nations framework convention on climate change: A commentary. *Yale Journal of International Law, 18*, 451–588.

Caponera, D. (1992). *Principles of water law and administration.* Rotterdam: Balkema.

Conti, K. I., & Gupta, J. (2014). Protected by pluralism? Grappling with multiple legal frameworks in groundwater governance. *Current Opinion in Environmental Sustainability, 11*, 39–47. doi:10.1016/j.cosust.2014.10.004

Conti, K. I., & Gupta, J. (2015). Global governance principles for the sustainable development of groundwater resources. *International Environmental Agreements: Politics, Law and Economics*, 1–23. doi:10.1007/s10784-015-9316-3

Dellapenna, J., & Gupta, J. (Eds.). (2009). *The evolution of the law and politics of water*. Dordrecht, Netherlands: Springer Netherlands.

Döll, P., Hoffmann-Dobrev, H., Portmann, F. T., Siebert, S., Eicker, A., Rodell, M., ... Scanlon, B. R. (2012). Impact of water withdrawals from groundwater and surface water on continental water storage variations. *Journal of Geodynamics, 59–60*, 143–156. doi:10.1016/j.jog.2011.05.001

Eckstein, G. E. (2007). Commentary on the UN international law commission's draft articles on the law of transboundary aquifers. *Colorado Journal of International Environmental Law & Policy, 18*(3), 537–610.

Eckstein, G. E., & Sindico, F. (2014). The law of transboundary aquifers: Many ways of going forward, but only one way of standing still. *Review of European, Comparative & International Environmental Law, 23*(1), 32–42. doi:10.1111/reel.12067

Gleeson, T., Befus, K. M., Jasechko, S., Luijendijk, E., & Cardenas, M. B. (2015, November). Modern groundwater. *Nature Geoscience*. doi:10.1038/NGEO2590

Green, T. R., Taniguchi, M., Kooi, H., Gurdak, J. J., Allen, D. M., Hiscock, K. M., ... Aureli, A. (2011). Beneath the surface of global change: Impacts of climate change on groundwater. *Journal of Hydrology, 405*(3–4), 532–560. doi:10.1016/j.jhydrol.2011.05.002

Gupta, J. (2014a). *The history of global climate governance*. Cambridge: Cambridge University Press.

Gupta, J. (2016). The watercourses convention, hydro-hegemony and transboundary water issues. *The International Spectator, 51*(3), 118–131. doi:10.1080/03932729.2016.1198558

Gupta, J. (2014b). Glocal politics of scale on environmental issues: Climate, water and forests. In F. J. G. Padt, P. F. M. Opdam, N. B. P. Polman, & C. J. A. M. Termeer (Eds.), *Scale-sensitive governance of the environment* (pp. 140–156). Oxford: John Wiley & Sons.

Gupta, J., Hildering, A., & Misiedjan, D. (2014). Indigenous people's right to water under international law: A legal pluralism perspective. *COSUST, 11*, 26–33.

Gupta, J. (2008). Global change: Analysing scale and scaling in environmental governance. In O. R. Young, H. Schroeder, & L. A. King (Eds.), *Institutions and environmental change: Principal findings, applications, and research frontiers* (pp. 225–258). Cambridge, MA: MIT Press.

Gupta, J., & Pahl-Wostl, C. (2013). Global water governance in the context of global and multilevel governance: Its need, form, and challenges. *Ecology and Society, 18*(4), 53. doi:10.5751/ES-05952-180453

Hanak, E., & Lund, J. (2012). Adapting California's water management to climate change. *Climatic Change, 111*, 17–44. doi:10.1007/s10584-011-0241-3

Heather, C., & Gleick, P. H. (2011). Climate-proofing transboundary water agreements. *Hydrological Sciences Journal, 56*, 711–718. doi:10.1080/02626667.2011.576651

Heuvelmans, G., Louwyck, A., & Lermytte, J. (2011). Distinguishing between management-induced and climatic trends in phreatic groundwater levels. *Journal of Hydrology, 411*(1–2), 108–119. doi:10.1016/j.jhydrol.2011.09.039

International Conference on Water and the Environment. (1992). *The Dublin Statement on Water and Sustainable Development*. ICWE. Retrieved 15 August 2017 from http://www.un-documents.net/h2o-dub.htm

International Law Association. (1966). *Helsinki rules on the uses of the water of international rivers*. In Report of the Fifty-Second Conference, p. 477. ILA. 15 August 2017 from http://www.internationalwaterlaw.org/documents/intldocs/Helsinki_Rules_with_comments.pdf

International Law Association. (1986). *Seoul rules on international groundwaters*. Report of the Sixty-Second Conference Held at Seoul. ILA. Retrieved 15 August 2017 from http://www.fao.org/docrep/008/y5739e/y5739e0h.htm

International Law Association. (2004). *Berlin rules on water resources.* In Fourth report of the Berlin conference on water resources (Annex IV). ILA. Retrieved 15 August 2017 from https://www.unece.org/fileadmin/DAM/env/water/meetings/legal_board/2010/annexes_groundwater_paper/Annex_IV_Berlin_Rules_on_Water_Resources_ILA.pdf

International Law Commission. (2008). *Draft articles on the law of transboundary aquifers.* Report of the International Law Commission on the Work of Its Sixtieth Session. UN Doc. A/63/10. ILC. Retrieved 15 August 2017 from http://legal.un.org/docs/?path=./ilc/texts/instruments/english/draft_articles/8_5_2008.pdf&lang=EF

Jiménez Cisneros, B. E., Oki, T., Arnell, N. W., Benito, G., Cogley, J. G., Döll, P., . . . Mwakalila, S. S. (2014). Freshwater resources. In *Climate change 2014: Impacts, adaptation, and Vulnerability. Part A: Global and sectoral aspects. Contribution of working group II to the fifth assessment report of the intergovernmental panel on climate change* (Vol. 1, pp. 229–269). Cambridge: Cambridge University Press.

Kaul, I., Conceicao, P., Le Goulven, K., & Mendoza, R. (2003). Why do global public goods matter today? In I. Kaul, P. Conceicao, K. Le Goulven, & R. Mendoza (Eds.), *Providing global public goods.* New York: Oxford University Press.

Klijn, A. M., Gupta, J., & Nijboer, A. (2009). Privatising environmental resources: The need for supervision. *Review of European Community and International Environmental Law, 18*(2), 172–184. doi:10.1111/j.1467-9388.2009.00639.x

Koskenniemi, M., & Leino, P. (2002). Fragmentation of international law? Postmodern anxieties. *Leiden Journal of International Law, 15*(3), 553–579. doi:10.1017/S0922156502000262

IPCC. (2007). Climate change 2007: Impacts, adaptation and vulnerability. In M. Parry, O. Canziani, J. Palutikof, P. van der Linden, & C. Hanson (Eds.), *Contribution of working group II to the fourth assessment report of the intergovernmental panel on climate change.* New York, NY: Cambridge University Press.

Margat, J., & van der Gun, J. (2013). *Groundwater around the world: A geographic synopsis.* Leiden, Netherlands: CRC Press.

McCaffrey, S. C. (2011). The International Law Commission's flawed draft articles on the law of transboundary aquifers: The way forward. *Water International, 36*(5), 566–572. doi:10.1080/02508060.2011.597094

McCaffrey, S. C. (2013). Keynote: sustainability and sovereignty in the 21st century. *Denver Journal of International Law and Policy, 41*(4), 507–514.

McIntyre, O. (2011). International water resources law and the international law commission draft articles on transboundary aquifers: A missed opportunity for cross-fertilisation? *International Community Law Review, 13*(3), 237–254. doi:10.1163/187197311X582386

Mekonnen, M. M., Gerbens-Leenes, P. W., & Hoekstra, A. Y. (2015). The consumptive water footprint of electricity and heat: A global assessment. *Environmental Science: Water Research & Technology, 1*(3), 285–297. doi:10.1039/C5EW00026B

Mirumachi, N. (2015). *Transboundary water politics in the developing world.* London: Routledge.

Nicholls, R. J., & Cazenave, A. (2010). Sea-level rise and its impact on coastal zones. *Science (New York, N.Y.), 328*(5985), 1517–1520. doi:10.1126/science.1185782

Nicol, A., & Cascão, A. E. (2011). Against the flow: New power dynamics and upstream mobilisation in the Nile Basin. *Review of African Political Economy, 38,* 317–325. doi:10.1080/03056244.2011.582767

Obani, P., & Gupta, J. (2014). Legal pluralism in the area of human rights: Water and sanitation. *Current Opinion in Environmental Sustainability, 11,* 63–70. doi:10.1016/j.cosust.2014.09.014

Pahl-Wostl, C., Gupta, J., & Bhaduri, A. (Eds.). (2016). *Handbook on water security.* Cheltenham, UK: Edward Elgar.

Pahl-Wostl, C., Gupta, J., & Petry, D. (2008). Governance and the global water system: A theoretical exploration. *Global Governance: A Review of Multilateralism and International Organizations, 14*(4), 419–435.

Ramsar Convention. (1971). *Ramsar convention on wetlands of international importance.* UN Treaty Series No. 14583.

Ranjan, P., Kazama, S., & Sawamoto, M. (2006). Effects of climate change on coastal fresh groundwater resources. *Global Environmental Change, 16*(4), 388–399. doi:10.1016/j.gloenvcha.2006.03.006

Siebert, S., Burke, J., Faures, J. M., Frenken, K., Hoogeveen, J., Döll, P., & Portmann, F. T. (2010). Groundwater use for irrigation – a global inventory. *Hydrology and Earth System Sciences Discussions, 7*(3), 3977–4021. doi:10.5194/hessd-7-3977-2010

Sindico, F. (2011). The Guarani aquifer system and the international law of transboundary aquifers. *International Community Law Review, 13*(3), 255–272. doi:10.1163/187197311X585338

Stephan, R. (2011). The draft articles on the law of transboundary aquifers: The process at the UN ILC. *International Community Law Review, 13*(3), 223–235. doi:10.1163/187197311X582287

Stoa, R. (2014). The United Nations watercourses convention on the dawn of entry into force. *Vanderbilt Journal of Transnational Law, 47*(5), 1321–1370.

Tamanaha, B. Z. (2008). Understanding legal pluralism: past to present, local to global. *Sydney Law Review, 30*, 375–411.

Treidel, H., Martin-Bordes, J. L., & Gurdak, J. J. (2011). *Climate change effects on groundwater resources: A global synthesis of findings and recommendations*. Leiden, Netherlands: CRC Press.

United Nations Conference on Environment and Development (1992). *Declaration of the United Nations conference on environment and development*. In Report of the United Nations Conference on Environment and Development. UN Doc A/Conf. 151/26 (Vol. 1). UNCED. Retrieved 15 August 2017 from http://www.un.org/documents/ga/conf151/aconf15126-1annex1.htm

United Nation Economic Commission for Europe. (1992). *Convention on the protection and use of transboundary watercourses and international lakes*. 1936 UNTS 269. UNECE. Retrieved 15 August 2017 from https://www.unece.org/fileadmin/DAM/env/water/pdf/watercon.pdf

United Nations General Assembly. (1994). United Nations convention to desertification in those countries experiencing serious drought and/or desertification, particularly in Africa. 1954 UNTS 3. UNGA. Retrieved 15 August 2017 from https://treaties.un.org/pages/ViewDetails.aspx?src=TREATY&mtdsg_no=XXVII-10&chapter=27&lang=en

United Nations General Assembly. (1997). United Nations convention on the law of the non-navigational uses of international watercourses. Registration No. 52106. UNGA. Retrieved 15 August 2017 from https://treaties.un.org/doc/Publication/UNTS/No%20Volume/52106/Part/I-52106-0800000280025697.pdf

United Nations General Assembly. (2000) *United Nations Millennium Declaration*. UN Doc. A/55/L.2. UNGA. Retrieved 15 August 2017 from http://www.un.org/millennium/declaration/ares552e.htm

United Nations General Assembly. (2008). Resolution on the law of transboundary aquifers. UN Doc. A/RES/63/124.UNGA. Retrieved 15 August 2017 from http://www.un.org/en/ga/search/view_doc.asp?symbol=A/RES/63/124

United Nations General Assembly. (2010). Resolution on human right to water and sanitation. UN Doc. A/64/292. UNGA. Retrieved December 29 2015 from http://www.un.org/News/Press/docs/2010/ga10967.doc.htm

United Nations General Assembly. (2015). Transforming our world: the 2030 agenda for sustainable development. UN Doc. A/res/70/1. UNGA. Retrieved 15 August 2017 from https://sustainabledevelopment.un.org/post2015/transformingourworld

UNHCR. (2010). *Resolution on human rights and access to safe drinking water and sanitation*. UN Doc. A /HRC/RES/15/9. United Nations Human Rights Council. Retrieved 15 August 2017 from http://ap.ohchr.org/documents/dpage_e.aspx?si=A/HRC/RES/15/9

Van Asselt, H. (2014). *The fragmentation of global climate governance: Consequences and management of regime interactions*. Oxford, UK: Edward Elgar.

Van Vliet, M., & Aerts, J. C. J. H. (2014). Adaptation to climate change in urban water management – flood management in the Rotterdam rijnmond area. In Q. Grafton, K. A.

Daniell, C. Nauges, J. D. Rinaudo, & N. W. W. Chan (Eds.), *Understanding and managing urban water in transition*. Dordrecht: Springer.

von Benda Beckmann, F. (2001). Legal pluralism and social justice in economic and political development. *IdS Bulletin, 32*(1), 46–56. doi:10.1111/j.1759-5436.2001.mp32001006.x

Vörösmarty, B. C. J., Hoekstra, A. Y., Bunn, S. E., Conway, D., & Gupta, J. (2015). What scale for water governance? *Science, 349*, 478–479. doi:10.1126/science.aac6009

Wada, Y., Van Beek, L. P. H., Sperna Weiland, F. C., Chao, B. F., Wu, Y. H., & Bierkens, M. F. P. (2012). Past and future contribution of global groundwater depletion to sea-level rise. *Geophysical Research Letters, 39*(9), 1–6. doi:10.1029/2012GL051230

Wada, Y., Wisser, D., & Bierkens, M. F. P. (2013). Global modeling of withdrawal, allocation and consumptive use of surface water and groundwater resources. *Earth System Dynamics Discussions, 4*, 355–392. doi:10.5194/esdd-4-355-2013

Zeitoun, M., & Allan, J. A. (2008). Applying hegemony and power theory to transboundary water analysis. *Water Policy, 10*(S2), 3–12. doi:10.2166/wp.2008.203

Zips, W., & Weilenmann, M. (2011). Introduction: Governance and legal pluralism – and emerging symbiotic relationship. In W. Zips & M. Weilenmann (Eds.), *The governance of legal pluralism: Empirical studies from Africa and beyond* (pp. 7–34). Vienna: LIT Verlag GmbG & CoKG.

Climate change considerations under international groundwater law

Raya Marina Stephan

ABSTRACT
Most of the earth's groundwater is in transboundary aquifers. This vital water resource will certainly be affected by climate change. This article reviews the global climate change framework to investigate how it considers water, and groundwater in particular. It then considers the international legal regime applicable to groundwater resources to explore how it deals with climate change and to what extent it is compatible with the UNFCCC framework. It concludes with identifying the limits and possibilities of the groundwater regime in addressing climate change.

Introduction

Groundwater represents about 98–99% of the available freshwater on earth; it is the earth's widest water reserve (Margat, 2008). It provides drinking water to at least 25% of the global population, with strong variation worldwide, exceeding 70% in Europe for example and reaching almost 100% in Libya and Saudi Arabia. Of the total abstracted groundwater, 65% goes for irrigation (Machard de Gramont et al., 2011). Worldwide, 2.5 billion people depend solely on groundwater resources to satisfy their basic daily water needs, mostly in arid and semi-arid regions where there is no surface water (WWAP, 2015). As water knows no boundaries, most of this groundwater is stored in aquifers spanning the territory of two or more states. So far, 592 transboundary aquifers have been identified, including 222 transboundary groundwater bodies as defined by the European Union Water Framework Directive (IGRAC, 2015).

Transboundary aquifers thus represent an important source of water on earth. While they were long neglected by international law, the situation has changed in recent years. The adoption in 2008 by the UN International Law Commission of the Draft Articles on the Law of Transboundary Aquifers (Draft Articles), subject of four resolutions of the UN General Assembly (annexed to two of them), has brought the issue up to the global level. The entry into force of the UN Convention on the Law of Non-navigational Uses of International Watercourses (1997) and the global opening of the Convention on the protection and use of transboundary watercourses and international lakes (1992) (UNECE Water Convention) have strengthened international water law, and certainly the rules applicable to transboundary aquifers. The adoption by the Meeting of the

Parties (Rome 2012) of the Model Provisions on Transboundary Groundwater,[1] inspired by the Draft Articles, confirms this tendency.

Because of the different issues at stake, water appears as a transversal element at the heart of the climate change, as a majority of its impacts are felt through alterations in the hydrological cycle. Since 2008, the IPCC experts have underlined the central role of water in their analysis of climate change, its impacts and possible measures (French Water Partnership, 2014). In its fifth report, the IPCC attributed to water a more prominent place. A whole chapter deals with 'freshwater resources'. Water is described as a cross-cutting issue in different parts related to concerns such as risks in coastal zones (storms and sea level rise), flooding zones, food security (droughts), wetlands, and heat waves (French Water Partnership, 2014). At the same time, the IPCC did not give proper consideration to the possible impacts of climate change on groundwater, concluding that it is still difficult to evaluate them and to distinguish them from human pressure. For the IPCC (Jiménez Cisneros et al., 2014), large uncertainties remain regarding the impact of climate change on groundwater. However, a change in the climate can have specific impacts on groundwater. If the climate is drier, with less rain, the recharge season might be shortened, leading to seasonal deficits in groundwater, and eventually to long-term decline in groundwater storage. Heavy rains could reverse the recharge deficit. But aquifers are recharged more effectively by prolonged steady rain, rather than short periods of intense rainfall. And heavy rains can increase the frequency and severity of groundwater-related floods. When properly managed, and not under stress, groundwater provides a unique buffer during extended dry periods. Aquifers can store water in a time of climate change.

This article deals with the consideration of climate change under the international legal regime for transboundary aquifers (UN Watercourses Convention, Draft Articles and UNECE Water Convention). It analyzes the frame and the mechanisms developed for water under the UN Framework Convention on Climate Change (UNFCCC). It then evaluates how international water law allows mitigation and adaptation measures at the level of a transboundary aquifer by encouraging and offering through its principles the necessary tools for developing cooperation among riparian states. The article checks the compatibility of this regime with the requirements of the UNFCCC and its related framework, to see how the different instruments can form a homogeneous set. It concludes by considering the possibilities and the limits of this regime in view of the climate change dimension.

Water under the international legal framework for climate change

The climate change legal framework

The development of the international legal framework for climate change started with the adoption of the UNFCCC at the Earth Summit in Rio de Janeiro in 1992. The Framework Convention entered into force on 21 March 1994, and today can be considered a universal convention, as it is ratified by 195 states. The convention was later 'operationalized' by the Kyoto Protocol (adopted 1997, entered into force 2005), which focuses on setting internationally binding emission reduction targets for greenhouse gases for developed states. A first commitment period started in 2008 and ended in 2012. The second commitment period defined at the 18th Conference of the Parties (COP) in Doha, Qatar, started in 2012 and is scheduled to end in 2020. For the post-2020 period a new universal agreement was discussed and adopted in Paris in December 2015 (COP 21). The agreement was open for ratification from 22 April 2016 to 21 April 2017. It entered into force on 4 November 2016, and is now (as of 15 July 2017) ratified by 153 parties.

The UNFCCC established a Conference of the Parties (Article 7), which meets yearly. The various COPs have adopted important decisions which have developed, created and clarified processes and mechanisms operating to meet the objectives of the UNFCCC and subsequent agreements. These decisions have created, *inter alia*, the Clean Development Mechanism, finance instruments (Green Climate Fund, Adaptation Fund, etc.), and the adaptation approach (Nairobi Work Program, Cancun Adaption Framework, etc.).

Objectives of the climate change legal framework

The first objective of the climate change framework represented by the UNFCCC is 'stabilization of greenhouse gas concentrations in the atmosphere at a level that would prevent dangerous anthropogenic interference with the climate system' (Article 2). This goal is reinforced in the Paris agreement, the aim of which is 'to strengthen the global response to the threat of climate change in the context of sustainable development and efforts to eradicate poverty' (Article 2).

These two instruments give water due regard, either directly or indirectly.

The indirect approach to water

The UNFCCC and the Paris agreement give indirect consideration to water. Both instruments include important provisions on two elements closely dependent on water resources: ecosystems and food production. The Paris agreement also refers to human rights, some of which are related to water.

Ecosystems. In Article 2, 'Objective', the UNFCCC adds that stabilization of greenhouse gas concentrations in the atmosphere 'should be achieved within a time frame sufficient to allow ecosystems to adapt naturally to climate change'. The convention acknowledges the possible effects on ecosystems, and their need for time to adapt in a natural way to climate change. This very prominent reference to ecosystems represents the first indirect approach to water in the UNFCCC. Water is a central element in any ecosystem. The value and the services ecosystems offer to humans have long been acknowledged. It is now recognized that ecosystems lie at the heart of the global water cycle, and that freshwater ultimately depends on the continued healthy functioning of ecosystems. Water availability and quality, in terms of direct use by humans, are also ecosystem services (WWAP, 2012). Therefore, protecting ecosystems means giving due regard to water resources. Conversely, healthy ecosystems rely on a certain quantity and quality of water for their proper functioning, and could therefore be affected by any change in the water cycle induced by climate change.

In the Glossary of Terms in the IPCC's Fifth Assessment Report (IPCC, 2014) an ecosystem is defined as 'a functional unit consisting of living organisms, their non-living environment, and the interactions within and between them'. It is clear from this definition that water is an essential component in an ecosystem, as it interacts with living organisms such as plants and animals. The definition does not provide a fixed delimitation of an ecosystem; the IPCC (2014) recognizes that the 'spatial boundaries' of an ecosystem 'depend on the purpose for which the ecosystem is defined'. So an ecosystem can be very wide, even covering the entire globe, or very small – a pond. In this sense the IPCC follows the Convention on Biological Diversity (also adopted at the Rio Summit in 1992). Under the Convention on Biological Diversity an ecosystem is not limited to 'any particular spatial unit or scale.... The term "ecosystem" ... can refer to any functioning unit at any scale. Indeed, the scale of analysis and action should be determined by the problem being addressed. It could, for example, be a grain of soil, a pond, a forest, a biome or the entire biosphere.'[2] Following these definitions, a surface water basin or an aquifer could be considered as an ecosystem, or only part of one. The UNFCCC gives here an indirect but nevertheless rather important consideration to water in its objectives. This consideration comes under the adaptation angle, as the convention (Article 2) mentions a necessary time frame for ecosystems *to adapt* naturally to the requested stabilization of greenhouse gas emissions.

Due regard to ecosystems also appears in the Paris agreement. The preamble already notes 'the importance of ensuring the integrity of all ecosystems', as well as 'the protection of biodiversity'. Further on, adaptation is recognized as important for the protection of ecosystems in view of the climate change (Article 7§2), and any adaptation action should take ecosystems into consideration (Article 7§5). While engaging in 'adaptation planning processes and the implementation of actions', the parties are requested to assess climate change impacts 'with a view to formulating nationally determined prioritized actions, taking into account vulnerable people, places and ecosystems' (Article 7§9c). The resilience of ecosystems to the adverse effects of climate change is identified as one of the 'areas of cooperation and facilitation to enhance understanding, action and support' in the frame of the loss and damage mechanism under the UNFCCC (Article 8§h).

Food production. According to Article 2 of the UNFCCC, the stabilization of greenhouse gas concentrations in the atmosphere should not threaten food production. In the Paris agreement, food security and protection receive high consideration. From the preamble, food production systems are recognized in the Paris agreement as particularly vulnerable to the adverse effects of climate change, and safeguarding food security is acknowledged as a fundamental priority. Under Article 2 of the agreement, food production needs to be protected while enhancing adaptation and resilience to climate change (§b). The same article closely links the need 'to strengthen the global response to the threat of climate change' to 'sustainable development and efforts to eradicate poverty'. Water is a central element in sustainable development and a driver for poverty eradication.

Reference to human rights. The Paris agreement mentions in its preamble various rights, such as human rights, the right to health and the right to development, which should be respected, promoted and considered 'when taking action to address climate change'. This set of rights obviously has a strong relation to water; the right to water is now a recognized human right (General Assembly, 2010), and the right to health includes the right to safe drinking water. The paragraph in the preamble referring to these rights echoes a paragraph in the preamble of Decision 1/CP21 adopting the agreement, showing the importance of the reference to human rights and the other mentioned rights as guiding principles in the implementation of the agreement.

Direct reference: water under adaptation

The only explicit mention of water in the UNFCCC is in Article 4.1§e. Article 4 relates to the 'Commitments' of the parties. Under Paragraph 1§e, the parties commit to 'cooperate in preparing for *adaptation* to the impacts of climate change; develop and elaborate appropriate and integrated plans for coastal zone management, *water resources* and agriculture, and for the protection and rehabilitation of areas, particularly in Africa, affected by drought and desertification, as well as floods'. Like ecosystems, water is considered under measures for adaptation to climate change. Water appears also as a transversal issue, as it is central for the other elements mentioned in Article 4.1§e, which are coastal zone management, agriculture, drought, desertification and floods. The reference to water in Article 4.1§e and to the other elements strongly linked to water seems to be a specific application of the requirement of the convention in relation to ecosystems. Under 1§f, Article 4 requires the parties to the convention to 'take climate change considerations into account, to the extent feasible, in their relevant social, economic and environmental policies and actions'. This provision implies for the parties to introduce considerations of climate change into their water policies, as part of their environmental policies. Under this same 1§f, the parties should ensure that any project or measure 'undertaken ... to mitigate or adapt to climate change' does not substantially affect the economy, the public health or the quality of the environment. This concern applies to the quality of water resources as part of public health, and of the environment.

The adaptation framework

Under the climate change regime, water clearly falls under adaptation, whether it is considered directly or considered through ecosystems, or as contributing to food production. Thus, any consideration of transboundary aquifers as an important source of water on earth under the climate change legal framework would fall under this key concept. The adaptation process has been developed progressively in the frame of the COPs, and it was confirmed recently in the Paris agreement.

The UNFCCC and the Paris agreement do not provide any definition of adaptation in the frame of the climate change regime. The Intergovernmental Panel on Climate Change (IPCC, 2014) defines adaptation as 'the process of adjustment to actual or expected climate and its effects. In human systems, adaptation seeks to moderate or avoid harm or exploit beneficial opportunities. In some natural systems, human intervention may facilitate adjustment to expected climate and its effects.'

The Least Developed Countries Work Programme adopted at COP 7 (2001) was perhaps the first initiative to include adaptation, but it concerned only a specific group of countries. Later, at COP 11 (Montreal, 2005), the Subsidiary Body for Scientific and Technological Advice was mandated to undertake a five-year project to address impacts, vulnerability and adaptation in relation to climate change, the Nairobi work programme. The Bali Action Plan (COP 13 2007) launched 'a comprehensive process to enable the full, effective and sustained implementation of the Convention through long-term cooperative action'. The plan is divided into five main categories: shared vision, mitigation, adaptation, technology and financing. The Cancun Adaptation Framework (CAF), adopted by Decision 1/CP16, was established at COP 16 (2010), in Cancun (Mexico). The parties affirmed that 'adaptation must be addressed with the same priority as mitigation' (§2b). Under this framework the COP invited 'all Parties to enhance action on adaptation' (§14) by 'planning, prioritizing and implementing adaptation actions, including projects and programmes' (§14a). These can be related to 'water resources; health; agriculture and food security; infrastructure; socio-economic activities; terrestrial, freshwater and marine ecosystems; and coastal zones', as clarified in a footnote in the decision. As in the UNFCCC, water is mentioned as a separate element, but the remaining components (health, agriculture, food, and also ecosystems and coastal zones) have a strong relationship with water.

The Paris agreement raised adaptation to the same level as mitigation, as it placed equal importance on reducing greenhouse emissions and adapting to climate change (Article 2§a,b), confirming the affirmation of the CAF.

The agreement provides numerous indications and details about adaptation, mainly in Article 7. It establishes adaptation as a global goal and challenge at all levels, from local to subnational, national, regional and international. Adaptation is recognized as important for the protection of people, livelihoods and ecosystems from the effects of climate change. Article 7, on adaptation, is strongly influenced by the CAF, to which it refers: 'Parties should strengthen their cooperation on enhancing action on adaptation, taking into account the Cancun Adaptation Framework' (Article 7§7). Various enactments of the CAF are formulated and confirmed by the Paris agreement, such as the provision that adaptation should be initiated by the countries, include gender considerations, and adopt a participatory approach. It should give due regard to vulnerable groups, communities and ecosystems and be based on the best available scientific knowledge, but also traditional and local knowledge when available, as provided in the CAF. International cooperation and support on adaptation efforts is recognized, especially for developing countries, and for actions such as

- sharing of information, good practices, experiences and lessons learned
- strengthening of institutions, to undertake the synthesis of relevant information and knowledge as well as technical support
- strengthening of the scientific knowledge on climate (research, systematic observation, early-warning systems).

It is expected that the parties will introduce adaptation into their national development, and environmental plans and policies.

According to the Paris agreement, each party should submit and periodically update an adaptation communication. This communication can be submitted in conjunction with any other document, including the national determined contribution. It will be recorded in a public registry managed by the secretariat of the convention. The Adaptation Committee is mandated to establish methodologies to assess adaptation needs, and in close cooperation with the Least Developed Countries Expert Group to propose recommendations to facilitate the mobilization of support for adaptation in developing countries (Decision 1/CP21).

Under the climate change global legal regime, water is considered under adaptation, whether directly, or indirectly through the mention of ecosystems, food security or human rights. Groundwater is an important source of water for ecosystems, for irrigation to ensure food security, and as drinking water. Most of it is found in transboundary aquifers. Therefore, under the climate change legal framework, transboundary aquifers are considered under the regime applied to water, which is adaptation. The adaptation process was defined throughout the various COPS, and confirmed in the Paris agreement. It includes, *inter alia*, cooperation between the parties, sharing of information, strong institutions and good scientific knowledge, with due consideration to ecosystems.

Climate change considerations under the legal framework for transboundary aquifers

The instruments of the legal framework for transboundary aquifers

As a water resource, transboundary aquifers fall under the adaptation regime of the climate change framework. Cooperation in adaptation measures among the riparian states of a transboundary aquifer is necessary to avoid the negative impacts of unilateral adaptation policies and plans and to assess and understand the effects of climate change on the resource as a whole, not only part of it. As mentioned earlier, transboundary aquifers fall under the scope of three different instruments: the UN Watercourses Convention (UNWC), the UNECE Water Convention and the Draft Articles.

The UN Watercourses Convention (1997)

The UNWC entered into force in 2014, and has now 36 parties. Its core principles are part of international customary law and therefore apply to all states. The convention 'applies to uses of international watercourses and of their waters for purposes other than navigation and to measures of protection, preservation and management related to the uses of those watercourses and their waters' (Article 1a, on Scope). It defines a watercourse as 'a system of surface waters and groundwaters constituting by virtue of their physical relationship a unitary whole and normally flowing into a common terminus' (Article 2§a). According to this definition, groundwaters fall within the scope of the UNWC *only when* they are related to a surface water body *and* they flow to the same terminus. In reality these conditions are not always met: even when an aquifer has a direct relationship with a surface water body it does not necessarily share the same terminus; it can discharge elsewhere. So the definition adopted by the UNWC leaves out a great number of transboundary aquifers (Stephan, 2011). Therefore, the convention's coverage of transboundary aquifers is limited. However it is considered in this article because of its global importance, and of the cases where its application will be ascertained.

The UNECE Water Convention

The Water Convention was originally a regional convention for the states of the UNECE region.[3] An amendment adopted in 2003, in force in 2015, opened the convention to all UN member states, making it a global convention. The convention applies to all transboundary waters (surface or groundwater) as long as they 'mark, cross or are located on boundaries between two or more States' (Article 1§1). It therefore has wide application regarding transboundary aquifers.

The Draft Articles on the Law of Transboundary Aquifers

The Draft Articles were prepared and adopted by the UN International Law Commission in 2008. The UN General Assembly adopted four resolutions on the topic: 63/124 (2008), 66/104 (2011), 68/118 (2011) and 71/150 (2016). The Draft Articles (DA) are annexed to two of them: Resolutions 63/124 and 68/118. A UN General Assembly resolution has no binding force. Resolution 63/124, as well as Resolution 66/104, '*encourages* the States concerned to make appropriate bilateral or regional arrangements for the proper management of their transboundary aquifers, taking into account the provisions of the draft articles'. In 2013 the language changes; the General Assembly seems to be putting a stronger emphasis on the Draft Articles, as it '*commends* to the attention of Governments the draft articles on the law of transboundary aquifers ... as guidance for bilateral or regional agreements and arrangements for the proper management of transboundary aquifers'. The General Assembly appears to express a will to promote the Draft Articles as guidelines. The Draft Articles cover the use of all transboundary aquifers / aquifer systems and their protection, preservation and management. They apply also to 'other activities' that could impact transboundary aquifers or aquifer systems, which means they extend beyond the application to the aquifer itself.

As mentioned above, transboundary aquifers fall under adaptation within the climate change framework. From the earlier definition (§I.2.3), it is clear that adaptation needs flexible principles, requiring the preservation and protection of the resource, and modifying its use patterns according to changing climatic conditions. Adaptation also requires observation; assessment of climate impacts and vulnerability; planning; implementation; and monitoring and evaluation of actions undertaken. The Paris agreement has added elements such as sharing of information, strengthening of institutions, and strengthening of the scientific knowledge on climate.

Therefore, the question is: does the international legal regime for transboundary aquifers allow any adaptation process between riparian states? Does it allow putting together the elements mentioned above? The consideration of adaptation under the legal regime for transboundary aquifers requires analysis of its key principles, and what possibilities they offer for adaptation.

The core principles of international water law

The three instruments codified the two core principles of international water law: equitable and reasonable use, and the no-harm rule. These two principles are strongly linked to a third principle deriving from general international law: the general obligation to cooperate.

Equitable and reasonable use

The principle of equitable and reasonable use implies equality of rights among the riparian states. Each aquifer state is entitled to use and benefit from the transboundary aquifer in an equitable and reasonable manner in its own territory, giving due consideration to the interests of the other riparians and to the adequate protection of the transboundary aquifer. It also implies for the aquifer states the duty to cooperate in the protection and development of the transboundary aquifer. The Draft Articles specifically mention the duty to 'establish individually or jointly a comprehensive utilization plan taking into account present and future needs of, and alternative water sources for, the aquifer States' (Article 4§c). The principle of equitable and reasonable use requires for its practical implementation a case-by-case assessment of factors. The UNWC and the Draft Articles provide an indicative list of factors with no priority assigned, except for the vital human needs (Article 5§2 DA, Article 10§2 UNWC). The two instruments require that 'special regard' be given to these needs, meaning that 'sufficient water to sustain life, including both drinking water and water required for production of food' (ILC, 2008) has to be provided. It has to be noted that the UNWC mentions climate in the list of factors (Article 6§1a).

Equitable and reasonable use is intrinsically a flexible principle, which requires a case-by-case assessment of the conditions in and around a transboundary aquifer for its implementation. Such an assessment could, and even in current times should, include climate conditions, their evolution, and their impact on the aquifer. It also requires the aquifer states to establish a utilization plan, either individually or jointly, giving due consideration to present and future needs, thus putting sustainable development at its heart.

The no-harm rule

The second core principle of international water law is the no-harm rule, which represents an obligation of conduct and not of result, in other words of due diligence, meaning that the harm is not caused intentionally or by neglect. In the course of their use of the transboundary aquifer within their territory, or when they are undertaking an activity which might have an impact on the aquifer, states are requested to take all appropriate measures to avoid any significant harm to other aquifer states, and also to states where the discharge zone is located.

The general obligation to cooperate

Another binding obligation on the riparian states of a transboundary aquifer under the three instruments derives from general international law, and is the general obligation to cooperate (Article 8 UNWC, Article 2§6 UNECE Water Convention, Article 7 DA). Cooperation between riparians is instrumental to full compliance with the other two obligations (UNECE, 2013). One of its practical applications is the regular exchange of data (Article 9 UNWC, Articles 6 and 13 UNECE Water Convention, Article 8 DA). The exchange of information is an element added by the Paris agreement regarding adaption.

While the no-harm rule does not appear as really relevant in the case of adaptation to climate change, equitable and reasonable use emerges as central, allowing flexibility and requesting that the concerned states consider various factors, whether related to the climate or not, and planning the use of the aquifers in a sustainable way. Sustainable development is also at the heart of the climate change legal framework formed by the UNFCCC and the Paris agreement.

Other provisions

Other provisions in the three instruments can be related to the elements of adaptation. They are the following.

Development of knowledge and exchange of information

Various provisions are related to the development of knowledge and exchange of information, such as the article on the regular exchange of data mentioned above, but also the articles on monitoring and research.

Monitoring

Monitoring is an essential element in the management of an aquifer. A permanent monitoring network allows 'to retrace and establish the evolution of the aquifer system over time' (Machard De Gramont et al., 2011, p. 44). In the case of adaptation to climate change, a monitoring network is also 'necessary for the regular update of the assessments, the scenarios of change, and the water balance projections in order to ensure flexible adaptation'.

Provisions on monitoring can be found in the UNECE Water Convention and in the Draft Articles. The Water Convention imposes a general obligation on the parties to 'establish programmes for monitoring the conditions of transboundary waters' (Article 4). In the frame of a bilateral or multilateral cooperation, the same convention obliges parties to 'establish and implement joint programmes for monitoring the conditions of transboundary waters' (Article 11). The Draft Articles require the aquifer states to monitor their transboundary aquifers, jointly if possible, and if necessary in collaboration with competent international organizations. They are also requested to identify 'key parameters' to monitor 'based on an agreed conceptual model of the aquifers or aquifer systems' (Article 13§2). Some key parameters are listed in Article 8, and they refer to geology, hydrogeology, hydrology and meteorology. This list is only indicative, and aquifer states can agree on other and additional parameters.

Research

The UNECE Water Convention includes two articles regarding cooperation on research. Article 5 relates to research on techniques for the prevention, control and reduction of transboundary impacts, giving indication on specific programmes. The article also provides that the states shall exchange the results. Article 12 focuses on 'specific research and development activities' for 'maintaining the water-quality objectives and criteria which they have agreed to set and adopt'.

Planning

Under the three instruments management is considered more than planning, which nevertheless is also mentioned. Management and planning are also strongly related to the establishment of joint institutions.

Management and planning

The three instruments include considerations about the management of a transboundary aquifer and the establishment of management plans (Article 24 UNWC, Articles 2§2b and 3§1i UNECE Water Convention, Article 14 DA). Under the UNWC and the Water Convention, management is promoted as sustainable, and as having the aim of 'rational and optimal utilization, protection and control of the watercourse' (Article 24 UNWC). The Draft Articles are more explicit on establishing a management plan, which is also mentioned in Article 4 on equitable and reasonable use, translating here the requirements for sound management of an aquifer.

Institutions: joint bodies/mechanisms

The requirement of establishing joint bodies/mechanisms for cooperation over and management of transboundary aquifers results from various provisions in the three instruments (Articles 8 and 24 UNWC, Article 9 UNECE Water Convention, and Articles 7 and 14 DA), with the very detailed tasks and responsibilities under the Water Convention, such as, *inter alia*, collecting data, elaborating joint monitoring programmes, and inventorying pollution sources.

Other useful provisions

Though not appearing under the elements of the definition of adaption, other provisions of the legal framework of transboundary aquifers present a useful relation to the climate change regime.

Ecosystems

While not related to the constitution of an adaptation process, ecosystems receive wide consideration under the legal framework for transboundary aquifers, as they do under the climate change regime. Groundwater is very important for ecosystems, which very often rely on it for 'the necessary humidity for many plant species, whether directly or indirectly through underlying humid soil. In arid environments, for instance, oases can only exist near shallow or resurgent groundwater' (Machard De Gramont et al., 2011, p. 16). Under the three instruments, aquifer states are under the obligation to protect and preserve the ecosystems depending on the aquifer (Article 20 UNWC, Article 10 DA and Articles 2 and 3 UNECE Water Convention). Under the Water Convention, even the restoration of an ecosystem is considered, as well as the application of the ecosystem approach.[4] The Draft Articles include among the factors relevant to equitable and reasonable use the role of the aquifer in the related ecosystem (Article 5§1i).

Cooperation with developing states

And finally, the Draft Articles include a provision on cooperation with developing states (Article 16) related to the scientific, educational, technical and legal aspects of the protection and management of transboundary aquifers, including, *inter alia*, capacity building, providing advice, and exchange of technical knowledge and experience. This provision is very useful in the case of adaptation, and climate issues generally, where developing states are experiencing the greatest impacts of climate change, though they are least prepared for them.

The above overview of the principles under the UNWC, the Water Convention and the Draft Articles shows that aquifer states can jointly consider and plan for adaptation measures. The rules deriving from these three instruments offer a rather comprehensive set of principles permitting the proper elaboration of adaptation strategies and policies for transboundary aquifers. The elements required for adaptation measures identified in the Paris agreement seem to be adequately addressed in these two instruments.

Climate change in specific agreements on transboundary aquifers

There are very few agreements on transboundary aquifers compared to the total number of transboundary aquifers. There are five agreements on the following transboundary aquifers:

- Genevese Aquifer (France, Switzerland), 1978/2008
- Nubian Sandstone Aquifer System (Chad, Egypt, Libya, Sudan), 1992
- North-Western Sahara Aquifer System (Algeria, Libya, Tunisia), 2008
- Guarani Aquifer System (Argentina, Brazil, Paraguay, Uruguay), 2010
- Al-Saq/Disi Aquifer (Jordan, Saudi-Arabia), 2015.[5]

These agreements do not deal with climate change as such. But they often offer possibilities for adaptation to circumstances that could be induced by climate change. For instance, under the current agreement on the Genevese Aquifer, the commission prepares an annual plan for the aquifer, giving due consideration as much as possible to the needs of the various users (Convention relative à la protection, à l'utilisation, à la réalimentation et au suivi de la nappe souterraine franco-suisse du Genevois, 2008, Article 2). The parties organize coordinated monitoring for the aquifer, for both water quantity and water quality. The agreement establishes a limit on abstractions on the French side, with the possibility of derogations following the opinion of the manager. The commission meets once a year, and more if needed, at the request of one of the parties. These provisions offer the flexibility to adapt to any change in the situation that could be induced by climate change. The organized monitoring of the aquifer gives an indication of its status. Cooperation on the Genevese Aquifer is now well established between the local water authorities on each side of the border, which is now used to adapt to any situation, and revise the adopted plan if necessary.

The Guarani agreement is built on the Draft Articles, and refers to its main principles, equitable and reasonable use and the no-harm rule (Acuerdo sobre el acuifero Guarani, 2010). It also insists on conservation of the environment. The agreement provides for exchange of data, development of joint projects and identification of critical areas. These principles were analyzed above in relation to climate change. The agreement is not yet in force, as so far it has been ratified by only two countries (Argentina and Uruguay) out of four.[6]

The agreement on the Nubian Sandstone Aquifer System establishes the Joint Authority for the Study and Development of the Nubian Sandstone Aquifer System (Constitution of the Joint Authority for the study and development of the Nubian

Sandstone Aquifer Waters, 1992). The agreement is an institutional agreement rather than a water management agreement. It provides the rules for the functioning of the Joint Authority. However, among its responsibilities, described in Article 3 of the agreement, the authority can collect data and information, and undertake studies on the aquifer system. It can also develop programmes and plans for the use of the water, on a scientific basis, and undertake to ration the consumption of water from the aquifer if necessary. The scientific aspect of these responsibilities offers grounds for the consideration of climate change in the development of programmes and plans. Furthermore, the countries and the Joint Authority in 2013 adopted and signed a Strategic Action Program (Regional Strategic Action Program for the Nubian Aquifer System, 2013), in which they recognized climate change as a challenge for consideration in the management of the aquifer system.

The agreement on the Al Saq/Disi Aquifer is about establishing a protected area on both sides of the border in which all extraction activities should be eliminated within five years, and a management area where the wells are exclusively for domestic use. The agreement creates a Joint Technical Committee in charge of implementing the agreement, supervising the status of the groundwater (quantity and quality) and the collection and exchange of information, statements and studies and their analysis. The agreement limits the extraction of water from the aquifer by creating the protected area and reserving the water from the management zone for domestic use. The responsibilities of the Joint Technical Committee are in line with the requirements under the adaptation process (above).

Finally, the ministerial declaration on the North-Western Sahara Aquifer System is about the creation of a consultation mechanism between the three riparian countries (Algeria, Libya and Tunisia) in view of mainly maintaining the common database. This instrument is certainly the one with the least consideration of climate change, not to say more.

Conclusion

After the adoption of the Paris agreement, COP 22 in Marrakech led to the adoption of the Marrakech Action Proclamation (Marrakech Action Proclamation for our Climate and Sustainable Development, 2016), which does not mention water but refers, like the UNFCCC and the Paris agreement, to poverty eradication, food security and the consideration of climate change in agriculture, which are all directly related to the availability of water. It also mentions the necessary adaptation efforts and support for the 2030 Agenda for Sustainable Development and its Sustainable Development Goals.

At the same time, the vast majority of the intended nationally determined contributions submitted by the states before COP 21 include a section on adaptation, or at least mention it.[7] Water, agriculture and health are the priority sectors identified for the adaptation (French Water Partnership & Coalition Eau, 2015). Water is thus without any doubt a central element in adaptation concerns; it is the most cited element, and it has strong links with agriculture and health.

Under the climate change regime, water receives more consideration than it seems at first reading. The UNFCCC mentions water directly in Article 4.1§e, related to adaptation. However, the UNFCCC and later the Paris agreement also give due regard to other elements relying heavily on water, such as ecosystems and food production. The consideration of water is therefore wider than the mention in Article 4.1§e of the UNFCCC. Under the UNFCCC, water, and therefore transboundary aquifers, fall under the adaptation regime, like ecosystems and food production. Adaptation is not defined as such, but the process has been conceptualized throughout the various COPs. This process includes the consideration of ecosystems, the best available scientific knowledge, the sharing of information and the strengthening of institutions, based on policies and plans.

The legal regime which applies to transboundary aquifers derives from three instruments: the UNWC, the Water Convention and the Draft Articles. Consideration of the management principles embodied in these instruments shows that cooperation among aquifer states in implementing an adaptation plan is in accord with the requirements of the UNFCCC and the Paris agreement. Indeed, the identified elements of the adaptation process were already embedded in the UNWC, the Water Convention and the Draft Articles.

Furthermore, the UNWC and the Draft Articles include provisions related to emergency situations (Article 28 UNWC, Article 17 DA), which cover natural causes such as floods, and which could represent an effect of climate change. Under such circumstances the affected State 'shall take all practicable measures necessitated by the circumstances to prevent, mitigate and eliminate any harmful effect of the emergency', in close cooperation with other states and with international organizations. The UNECE Water Convention includes an article on 'Warning and Alarm Systems' (Article 14), under which the riparian states have the obligation to 'inform each other about any critical situation that may have transboundary impact', and to set up warning and alarm systems.

These provisions are not directly linked to establishing an adaptation strategy and plan for a transboundary aquifer, but could be very useful in case of serious effects of climate change. The same conclusion can be applied to the specific agreements on transboundary aquifers.

Notes

1. https://www.unece.org/fileadmin/DAM/env/water/publications/WAT_model_provisions/ece_mp.wat_40_eng.pdf, accessed 15 April 2016.
2. https://www.cbd.int/ecosystem/description.shtml, accessed 25 February 2016.
3. The UNECE covers Europe, Central Asia, Israel, Canada and the USA.
4. Under the CBD, the ecosystem approach is defined as a 'strategy for the integrated management of land, water and living resources that promotes conservation and sustainable use in an equitable way' (Secretariat of the Convention on Biological Diversity, 2004).
5. An agreement on the Iullemeden aquifer system (Mali, Niger and Nigeria) was prepared, but nothing happened afterwards. A protocol on the wider transboundary aquifer system (Iullemeden and Taoudeni/Tanezrouft) is under circulation between the seven riparian countries (Algeria, Benin, Burkina-Faso, Mali, Mauritania, Niger and Nigeria).
6. On 2 May 2017, the Brazilian Senate approved the Guarani Aquifer Agreement, paving the way for its ratification.
7. As of 8 November 2015, 129 INDC were available. Out of these, 106 (82%) include or mention adaptation.

Disclosure statement

No potential conflict of interest was reported by the author.

References

Acuerdo sobre el acuifero Guarani. (2010). Retrieved from https://legislativo.parlamento.gub.uy/temporales/S2012030486-009260996.pdf.

Agreement between the Government of the Hashemite Kingdom of Jordan and the Government of the Kingdom of Saudi Arabia for the Management and Utilization of the Ground Waters in the Al-Sag/Al-Disi Layer. 2015. Retrived from http://programme.worldwaterweek.org/sites/default/files/disi_aquifer_agreement-english2015.pdf

Constitution of the Joint Authority for the study and development of the Nubian Sandstone Aquifer Waters. (1992). (English translation). Retrieved from http://www-naweb.iaea.org/napc/ih/documents/Nubian/Nubian_final_MSP_Sandstone.pdf

Convention relative à la protection, à l'utilisation, à la réalimentation et au suivi de la nappe souterraine franco-suisse du Genevois. (2008). Retrieved from https://www.ge.ch/legislation/accords/doc/3038.pdf

French Water Partnership. (2014). Enseignements du GIEC: L'adaptation du secteur de l'eau aux changements globaux et climatiques. Retrived from http://www.partenariat-francais-eau.fr/wp-content/uploads/2015/06/2015-02-12-Enseignements-du-GIEC.pdf

French Water Partnership & Coalition Eau. (2015). Note d'analyse concernant la prise en compte de l'eau dans les contributions nationales determines par pays de la COP 21. Retrieved from http://www.partenariat-francais-eau.fr/wp-content/uploads/2015/11/2015-11-27_Note-danalyse-Eau-dans-les-INDC.pdf

General Assembly. (2010). *Resolution 64/292 the human right to water and sanitation.* New York: United Nations.

IGRAC. (2015). *Transboundary aquifers of the world.* Delft: IGRAC. Retrieved from https://www.un-igrac.org/sites/default/files/resources/files/TBAmap_2015.pdf

ILC. (2008). Report of its sixtieth session, A/63/10. Retrieved from http://legal.un.org/docs/?symbol=A/63/10(SUPP)

IPCC. (2014). Annex II: Glossary. In J. Agard, E. L. F. Schipper, J. Birkmann, M. Campos, C. Dubeux, Y. Nojiri, … L. L. White (Eds.), *Climate change 2014: Impacts, adaptation, and vulnerability. Part B: regional aspects. Contribution of working group II to the fifth assessment report of the intergovernmental panel on climate change.* New York, NY: Cambridge University Press.

Jiménez Cisneros, B. E., Oki, T., Arnell, N. W., Benito, G., Cogley, J. G., Döll, P., … Mwakalila S. S. (2014). Freshwater resources. In C. B. Field, V. R. Barros, D. J. Dokken, M. D. Mach, T. E. Mastrandrea, M. Bilir, … M. Chatterjee (Eds.), *Climate change 2014: Impacts, adaptation, and vulnerability. Part A: global and sectoral aspects. Contribution of working group II to the fifth assessment report of the intergovernmental panel on climate change.* New York, NY: Cambridge University Press.

Machard de Gramont, H., Noël, C., Oliver, J.-L., Pennequin, D., Rama, M., & Stephan, R. M. (2011). *Towards a joint management of transboundary aquifer systems, A savoir n° 3.* Paris: Agence Française de Développement.

Margat, J. (2008). *Les eaux souterraines dans le monde.* Paris: BRGM, UNESCO.

Marrakech Action Proclamation for our Climate and Sustainable Development. (2016). Retrieved from http://unfccc.int/files/meetings/marrakech_nov_2016/application/pdf/marrakech_action_proclamation.pdf

Regional Strategic Action Program for the Nubian Aquifer System. (2013). IAEA, UNDP/GEF. Retrieved from https://www.iaea.org/sites/default/files/sap180913.pdf

Secretariat of the Convention on Biological Diversity. (2004). *The ecosystem approach, (CBD Guidelines).* Montreal, QC: Secretariat of the Convention on Biological Diversity.

Stephan, R. M. (2011). The draft articles on the law of transboundary aquifers: The process at the UN ILC. *International Community Law Review*, *13*, 223–235. doi:10.1163/187197311X582287

UNECE. (2013). *Guide to implementing the water convention*. New York, NY: United Nations. Retrieved from https://www.unece.org/fileadmin/DAM/env/water/publications/WAT_Guide_to_implementing_Convention/ECE_MP.WAT_39_Guide_to_implementing_water_convention_small_size_ENG.pdf

WWAP (World Water Assessment Programme). (2012). The United Nations world water development report 4: Managing water under uncertainty and risk. Paris: UNESCO.

WWAP (World Water Assessment Programme). (2015). The United Nations world water development report 5, water for a sustainble world facts and figures. Paris: UNESCO.

Update to Chapter 9: Climate change considerations under international groundwater law

Raya Marina Stephan

At COP 23 (Bonn, 6-17 November 2017), the Parties to the UNFCCC adopted a final declaration, the Fiji Momentum for Implementation. The declaration acknowledges the progress made in the work programme to implement the Paris agreement, which is expected to be completed at COP 24 (November 2018), without mentioning any of its elements. It announces the launch of a facilitative dialogue (the Talanoa dialogue) in January 2018, intended to be solution oriented and conducted in a cooperative manner. And finally the declaration stresses the importance of the pre-2020 implementation.

None of these initiatives has given a practical indication of the consideration by the UNFCCC or the Paris Agreement of factors related to water, such as ecosystems, food production or human rights.

Index